BUILDING A NATION

Good Governance, Trust Building And Democratization In Sierra Leone

Dr. Robert B. Kargbo, Dr. Alfred Veenod Fullah, Mohamed B. J. Jamboria, Hassanatu Turay, Alieu M. Bah, Marian Kamara, Dr. Mohamed Sesay, Fatima Lydia Sessay, Zainab Melvina Omoyinmi, Fatima Sesay, Hassan A. Turay, Prof. M. Y. Bangura, Foday John Turay, Abdul Fonti, Mariatu Kamara, Marie H. Kamara

Building A Nation

Copyright © 2024

All rights reserved. No portion of this book may be reproduced or stored in a retrieval system or translated in any form or by any means – electronic, mechanic, photocopy, recording, scanning, or other – except for brief quotations in critical reviews or articles, without the prior written permission of the author.

TABLE OF CONTENTS

CHAPTER 1: AWAKENING TO THE NEED FOR CHANGE 4

CHAPTER 2: THE POWER OF OUR STORY 21

CHAPTER 3: ACTIONABLE BLUEPRINT FOR GOVERNANCE REFORM AND DEVELOPMENT 40

CHAPTER 4: ESTABLISHING THE FRAMEWORK 59

CHAPTER 5: PRINCIPLES OF GOOD GOVERNANCE 71

CHAPTER 6: KEY STRATEGIES FOR ENHANCING GOVERNANCE 98

CHAPTER 7: GUIDE RAILS FOR UPHOLDING DEMOCRATIC PRINCIPLES IN SIERRA LEONE 123

CHAPTER 8: DETRIMENTAL PRINCIPLES TO DEMOCRATIC GOVERNANCE IN SIERRA LEONE 130

CHAPTER 9: STRATEGIES FOR COMBATING DETRIMENTAL PRINCIPLES 177

CHAPTER 10: THE SIERRA LEONE ADVOCACY MOVEMENT 349

CHAPTER 11: OVERALL CONCLUSION AND CALL TO ACTION 392

ACKNOWLEDGMENTS 397

ABOUT THE AUTHORS 399

CHAPTER 1:
AWAKENING TO THE NEED FOR CHANGE

Opening Fact or Assumption: Did you know that despite their deep desire for progress, many Sierra Leoneans in the diaspora feel disconnected from the political and economic changes they wish to see back home? This sense of detachment often stems from the lack of effective channels to influence policies and contribute meaningfully to national development. As a result, their efforts seem scattered and ineffectual, leaving them frustrated and yearning for a tangible impact.

Further, many Sierra Leoneans at home remain caught in a quagmire, experiencing a social and psychological disconnect that fosters perceptions often undermining efforts to establish ethical standards. This disconnect hinders collective participation in governance, trust-building, and the democratization of the country—a challenge that has persisted since independence in 1961.

Thus, despite efforts to drive change, a persistent psychological tendency to prioritize personal interests over institutional considerations has posed significant challenges. Consequently, concepts like nationhood, patriotism, and collective aspirations vary widely among individuals, often hindering meaningful nation-building efforts.

Supporting Materials: To better understand this pervasive issue, consider the significant role remittances play in Sierra Leone's economy. According to the World Bank, remittances from the Sierra Leonean diaspora amounted to nearly $300 million in 2020, constituting about 13% of the country's GDP.[1] This financial support is crucial for many families, covering expenses like education, healthcare, and housing. However, despite this substantial economic contribution, there is a stark disconnect between these financial inflows and the diaspora's ability to influence governance and development policies.

Quote from a Prominent Figure: Dr. Akinwumi Adesina, President of the African Development Bank, pointed out that "the African diaspora has become the largest financier of Africa! And it is not debt; it is 100% gifts or grants, a new form of concessional financing that is the key to livelihood security for millions of Africans."[2] This statement underscores the immense potential of the diaspora's financial contributions and the need for structured engagement to maximize their impact.

"The African Diaspora in the United States is a source of strength. It encompasses African Americans, including descendants of enslaved Africans, and nearly two million African immigrants who have close familial, social, and economic connections to the continent. The African Diaspora—i.e., people of native African origin living outside the continent—has been described as the sixth region of African Union." - Vice President Kamala Harris.

---

[1] (https://data.worldbank.org/indicator/BX.TRF.PWKR.CD.DT?locations=SL, accessed 6/16/2024).

[2] (https://www.afdb.org/en/news-and-events/press-releases/diasporas-remittances-investment-and-expertise-vital-africas-future-growth-say-participants-african-development-bank-forum-57024, accessed 6/16/2024).

This quote was delivered by Vice President Kamala Harris during the African and Diaspora Young Leaders Forum, held in December 2022, as part of the U.S.-Africa Leaders' Summit.[3] The forum reflected the Biden-Harris Administration's commitment to strengthening the dialogue between U.S. officials and the African Diaspora. This initiative underscores the administration's recognition of the significant contributions of the African Diaspora to America's growth and prosperity and aims to enhance cultural, social, political, and economic ties between the U.S. and Africa.

The Disconnect Between Financial Contribution and Political Influence: The data paints a clear picture: despite their substantial financial reparations and other contributions, the diaspora's impact on Sierra Leone's political and economic landscape is minimal due to an exclusion from direct participation in politics and policy making. This disconnection not only hampers potential progress but also fosters a sense of frustration and helplessness among those eager to see tangible change.

Many diaspora members find themselves questioning the efficacy of their contributions. They wonder why, despite sending money regularly, advocating for positive change and even initiating projects for social welfare that normally must be undertaken by the state, there is little to show in terms of political stability, economic and social development. This frustration often leads to disengagement, with individuals feeling that their efforts are in vain.

---

[3] (https://www.presidency.ucsb.edu/documents/fact-sheet-us-africa-partnership-elevating-diaspora-engagement, accessed 6/16/2024).

In Sierra Leone, there is a degree of mistrust toward diasporan Sierra Leoneans, largely stemming from negative experiences with a few individuals who have returned and attempted to reintegrate into society. Many who have repatriated and directly entered politics have not set the best ethical examples. This is evident in both the previous and current regimes, which are predominantly composed of individuals who once lived in the diaspora before returning to engage in governance.

Real-Life Example: Consider the situation of many Sierra Leoneans in the diaspora. They send significant portions of their earnings back home each month, hoping to support their families and contribute to community projects. Despite their best efforts, they often feel disconnected from the actual impact of their contributions. Decisions about how the funds are used are frequently made without their input, leaving them uncertain about whether their support leads to sustainable development. This sense of powerlessness is a common experience among the diaspora, who feel that their voices and efforts are not adequately valued in the national discourse.

This scenario underscores a broader issue faced by many diaspora members, highlighting the need for more structured engagement and transparency to ensure their contributions can drive meaningful change.

On the other hand, a few individuals who returned and aligned themselves with poor governance regimes have demonstrated ethical shortcomings. This behavior has deepened mistrust toward diasporan intentions and their aspirations for change—a sentiment that has long been a significant barrier to development.[4]

Nunn et al (2011) states *"Combining contemporary individual-level survey data with historical data on slave shipments by ethnic group, we ask whether the slave trade caused a culture of mistrust to develop within Africa. Initially, slaves were captured primarily through state organized raids and warfare, but as the trade progressed, the environment of ubiquitous insecurity caused individuals to turn on others—including friends and family members—and to kidnap, trick, and sell each other into slavery (Sigismund Wilhelm Koelle 1854; P. E. H. Hair 1965; Charles Piot 1996). We hypothesize that in this environment, a culture of mistrust may have evolved, which may persist to this day."*

This culture still has a deep impact of trust and interpersonal relations even within family circles.

Conclusion of the Introduction: The challenge, therefore, lies in building trust and bridging the polarization caused by mistrust and its social and psychological effects. How can the diaspora transition from being mere financial contributors to becoming influential stakeholders in Sierra Leone's development? Another question is how can trust between home based and those in the diaspora be built to foster nation building? The answer lies in creating structured, effective channels for education, engagement and participation. This book aims to provide the roadmap for achieving that transformation, guiding you from frustration to fulfillment, from scattered efforts to strategic

---

[4] https://www.imf.org/en/Blogs/Articles/2024/01/25/how-distrust-of-government-by-marginalized-people-fuels-conflict-in-africa

influence.

As we delve deeper into this journey, we will explore practical steps and real-world strategies to fully harness the diaspora's potential and how a sustainable synchronisation of social psychologies shall be instituted to produce the desired results. Together, we can turn dreams of progress into a tangible reality, ensuring that every effort counts towards building a brighter future for Sierra Leone.

Connecting with Our Audience: In a recent conversation with Fatmata, a Sierra Leonean entrepreneur living in London, she expressed her frustration at being unable to influence the political climate back home. Despite her financial contributions and deep love for her country, she felt her voice was unheard. This sentiment is shared by many in the diaspora, who, like Fatmata, are eager for change and want to be part of the solution. They seek immediate and effective ways to improve the political and economic conditions in Sierra Leone, striving for better living standards and a brighter future for all.

Real-Life Example: Consider the challenges faced by many Sierra Leoneans in the diaspora. For instance, members of the Sierra Leonean community in Dallas-Fort Worth, Texas, have exemplified unity and collective effort through the Association of Sierra Leonean Organizations in Texas (ASLOT)[5] ASLOT aims to mobilize resources and foster development in healthcare, education, infrastructure, and social empowerment in Sierra Leone. Despite these efforts, many diaspora members feel their contributions are not adequately recognized or utilized.

---

[5] (https://cocorioko.net/sierra-leoneans-in-the-diaspora-faces-challenges-dilemma-and-possibilities/accessed 6/15/2024).

This feeling is further echoed in broader diaspora communities. The DiasporaEngager platform, for example, aims to connect Sierra Leoneans globally, fostering collaboration between those abroad and those in Sierra Leone. The platform's mission is to detect, harvest, and transfer resources, services, and opportunities to bridge the developmental gaps between the diaspora and their homeland.[6]

Understanding Our Intended Writers Aspirations: These writers envision themselves as agents of change, contributing to a thriving, democratic, and just society. They are motivated by a desire to see doable and sustainable solutions to current challenges and are ready to act. By organizing themselves and leveraging their collective strengths, such as through platforms like Diaspora Engager, they seek to mobilize resources, share knowledge, and influence decision-making processes in Sierra Leone.

Many in the diaspora are inspired by successful models from other countries. For instance, Ghana's "Joseph Project" invited its diaspora to reconnect with their homeland and actively participate in its development. This initiative created avenues for meaningful engagement and national development. Similarly, Nigeria's Nigerians in Diaspora Organization (NIDO) has provided a structured framework for diaspora engagement, allowing Nigerians abroad to contribute to their country's progress without necessarily returning home.

Conclusion of the Section: By connecting with our reader's

---

[6] (https://www.diasporaengager.com/extPage/SierraLeoneanDiasporaPlatform, accessed 6/16/2024).

identity and highlighting real-life examples and aspirations, we have illustrated the importance of structured engagement and collective action. Our ideal reader is someone who not only wishes to see change but is also ready to be part of the solution. This narrative will resonate with Sierra Leoneans in the diaspora who are eager to contribute meaningfully to their homeland's development.

By fostering unity and creating structured engagement channels, the Sierra Leonean diaspora can significantly impact the country's political and economic landscape, ensuring that their efforts lead to sustainable and meaningful progress. Through collaborative platforms and inspired by successful models from other countries, Sierra Leoneans abroad can bridge the gap between aspiration and action, transforming their homeland for the better.

The Cost of Inaction

Consequences of Not Solving the Problem: The inability to influence significant change back home carries a heavy toll. It perpetuates a cycle of frustration and disengagement, where the diaspora's potential remains untapped. This lack of effective participation not only affects personal morale but also deprives Sierra Leone of valuable insights and resources. Decisions are made without inclusive consultations, and policies are often implemented without the necessary public support, leading to ineffective governance and missed opportunities for progress.

Illustrative Examples: The consequences of inaction are severe and multifaceted. When the diaspora is not effectively engaged, their contributions can be sporadic and uncoordinated, leading to suboptimal outcomes.

1. **Dormant Investments**: Many potential investments from the diaspora remain untapped due to a lack of structured engagement and transparency. For example, the African Development Bank highlighted that diaspora remittances, while substantial, are often used for immediate family needs rather than long-term investments or development projects. This is a missed opportunity for significant economic impact. Without proper channels, the funds that could stimulate local businesses, create jobs, and foster economic growth lie dormant.

2. **Socio-Economic Development Lags**: Socio-economic development lags when the diaspora's expertise and resources are not fully utilized. Countries like Ghana and Nigeria have shown that structured diaspora engagement can lead to substantial national benefits. Ghana's "Joseph Project" and Nigeria's NIDO have successfully mobilized their diasporas to contribute to national development through investments and professional expertise. In contrast, the lack of similar initiatives in Sierra Leone means that the country misses out on the developmental benefits that an engaged and organized diaspora can bring.

3. **Missed Opportunities for Progress**: Decisions made

without inclusive consultations often result in policies that lack broad support and fail to address the needs of the population effectively. For instance, during discussions at the African Development Bank Forum, it was noted that the exclusion of diaspora voices leads to policies that do not reflect the ground realities or the aspirations of the people. This disconnection results in ineffective governance and hinders progress.

4. **Frustration and Disengagement**: The lack of impact can lead to frustration and eventual disengagement among diaspora members. For example, Sierra Leoneans in the diaspora have expressed feelings of helplessness when their efforts to contribute to national development are not recognized or utilized effectively. This disengagement not only affects their morale but also reduces the potential influx of resources and ideas that could drive the country's development.

Hypothetical Scenario: Imagine a scenario where a group of Sierra Leonean doctors abroad wants to set up a telemedicine network to improve healthcare access in rural Sierra Leone. Without structured support and clear communication channels with the government, their initiative stalls. Bureaucratic red tape, lack of local cooperation, and insufficient infrastructure turn what could have been a transformative project into a failed attempt. Meanwhile, communities continue to suffer from inadequate healthcare, and the doctors feel their efforts are wasted.

Conclusion of the Section: The cost of inaction is high, affecting both the diaspora and the home country. The potential for progress remains unrealized, and frustration grows among those who wish to contribute. By recognizing these consequences and working towards inclusive and structured engagement, Sierra Leone can harness the full potential of its diaspora, ensuring that their contributions lead to meaningful and sustainable development.

The Benefits of Solving the Problem

Imagining a Better Future: Imagine a scenario where the diaspora is fully engaged and effectively influencing the political and economic landscape of Sierra Leone. Envision a country where policies are crafted with inclusive consultations, reflecting the collective aspirations of all Sierra Leoneans. This transformation would lead to improved living conditions, robust economic growth, and a thriving democratic society. For the diaspora, this means not only seeing tangible results from their contributions but also gaining a profound sense of fulfillment and connection to their homeland.

When the home-based elites and diaspora engage in a trust drive collaborative discuss and initiatives that is mutually meaningfully, the respective contributions can significantly impact Sierra Leone's development. Consider the potential of fully harnessed diaspora engagement: policies would be more reflective of the population's needs, ensuring greater public support and effective implementation. This inclusive approach can foster unity, reduce political tension, and create a more stable and prosperous nation.

Specific Benefits

1. **More Stable Political Environment**: Engaging the diaspora can lead to a more inclusive and stable political environment. With their input, policies are more likely to address the needs and concerns of a broader spectrum of society, reducing disenfranchisement and fostering national unity. For example, Ghana's "Joseph Project," aimed at reconnecting the diaspora with the homeland, has successfully fostered a sense of belonging and participation among Ghanaians abroad, leading to a more cohesive national identity.

2. **Increased Foreign Investments**: A well-engaged diaspora can attract substantial foreign investments. Diaspora members often act as bridges between their home and host countries, facilitating business partnerships and investment opportunities. Countries like India and China have successfully leveraged their diasporas to attract investments and promote economic growth. Sierra Leone could experience similar benefits by establishing robust frameworks for diaspora engagement, leading to increased foreign direct investments and economic expansion.

3. **Empowerment of Citizens**: Empowering the diaspora to participate in national development can have a transformative effect on local communities. Projects led by diaspora members, such as healthcare initiatives, educational programs, and infrastructure developments, can significantly improve living standards.

For instance, the Nigerian diaspora's contributions

through NIDO have led to advancements in various sectors, including education and healthcare, demonstrating the potential impact of empowered diaspora communities.

4. **Improved Education and Healthcare**: Engaging the diaspora can lead to substantial improvements in formal and non-formal education and healthcare. Many diaspora members possess advanced skills and knowledge that can be transferred to their home country through training programs, investments in educational institutions, and healthcare projects. This can elevate the quality of education and healthcare services, ultimately improving the overall quality of life for Sierra Leoneans.

5. **Overall Quality of Life**: The ripple effect of increased political stability, foreign investments, and improved trusted public services leads to a higher overall quality of life. With better education, healthcare, and economic opportunities, citizens can achieve their full potential, contributing to a more prosperous and vibrant society. For the diaspora, seeing these tangible results reinforces their connection to their homeland and validates their efforts, creating a virtuous cycle of engagement and development.

Conclusion of the Section: By solving the problem of diaspora disconnection, Sierra Leone stands to gain immensely. A fully engaged diaspora can drive political stability, economic growth, and social development, transforming the country into a beacon of progress and democracy.

The benefits of this engagement are far-reaching, impacting not only the diaspora and their immediate families but also the broader national community. This vision of a united, prosperous Sierra Leone can become a reality through strategic and inclusive diaspora engagement, ensuring that every contribution counts towards building a brighter future.

In the next chapter, we will share our personal journey and the strategies that may help us overcome these challenges. You'll learn how to harness your potential and become a key player in Sierra Leone's transformation. Together, we will explore practical steps to bridge the gap between aspiration and action, ensuring that our efforts lead to meaningful and lasting change. Stay with us as we embark on this transformative journey towards a better Sierra Leone.

As we move forward, it's essential to understand that transformation begins with each one of us. By examining real-life examples and proven strategies, we can draw valuable lessons on how to effectively contribute to national development. Our journey, and those of many others, serves as a testament to the power of determination and collective effort. Through practical insights and actionable steps, the upcoming chapter aims to equip you with the tools necessary to make a significant impact.

You'll discover how to leverage your skills, resources, and networks to foster development in Sierra Leone. Whether through investing in local businesses, participating in community projects, or advocating for policy changes, there are numerous ways to make your contributions count.

We will delve into specific strategies that have been successful in other contexts and explore how they can be adapted to the Sierra Leonean landscape.

Motivational Ending: Remember, the journey of a thousand miles begins with a single step. As we embark on this transformative journey together, your commitment and passion are the catalysts for change. Each action you take, no matter how small, contributes to the broader goal of a united, prosperous Sierra Leone. Your efforts are not in vain; they are the building blocks of a brighter future.

As Nelson Mandela once said, 'It always seems impossible until it's done.' Let's make the impossible possible, one step at a time. In the next chapter, we will guide you through the practical steps to harness your potential and turn your aspirations into reality. Together, we can achieve lasting and meaningful change for our beloved Sierra Leone.

By setting this motivational tone, we aim to inspire and empower readers, encouraging them to see themselves as integral parts of Sierra Leone's transformation. The journey ahead is challenging, but with determination and collective effort, we can overcome these challenges and build a better future for all.

Reflection and Engagement Questions: This chapter introduces the readers to the critical need for change in Sierra Leone, highlighting the importance of engagement from the local and diaspora community. It emphasizes the collective responsibility in addressing governance issues and the benefits of a unified approach. The chapter aims to resonate with readers by identifying common aspirations and challenges, thereby fostering a sense of connection and urgency.

Reflection Questions

1. **Understanding Connection:**
   - How do you personally identify with the issues discussed in this chapter regarding the disconnect between the diaspora and the political and economic developments in Sierra Leone?

   - _____
     _____

2. **Evaluating Impact:**
   - In what ways do you think the financial contributions from the diaspora, such as remittances, can be more effectively utilized to influence positive change in Sierra Leone?

   - _____
     _____

3. **Addressing Challenges:**
   - What are the primary obstacles that prevent effective diaspora engagement in Sierra Leone's governance, and how can these be overcome?

   - _____
     _____

4. **Imagining Solutions:**
   - Reflect on a time when you felt disconnected from a significant cause.

     What steps did you take to bridge that gap, and how can similar strategies be applied to enhance diaspora engagement in Sierra Leone?

- _____
_____

5. **Personal Commitment:**

    - Considering the need for a collective effort, what role do you envision for yourself in contributing to the governance and development of Sierra Leone?

    - _____
    _____

By answering these questions, readers can internalize the chapter's content, reflect on their personal connections to the issues, and consider actionable steps they can take to contribute to Sierra Leone's progress.

CHAPTER 2:
THE POWER OF OUR STORY

Start with the End in Mind: We can imagine the day we finally walked into our boss's office and handed in our resignation letter to pursue our dream of contributing to Sierra Leone's development full time. The sense of fulfillment and purpose that washes over us could be indescribable. But before we reached this milestone, we embarked on a journey filled with challenges and frustrations. Let us take you back to the beginning, to the moments that shaped our path and brought us here. If you permit, we'd love to share some of the obstacles we faced and how they fueled our determination to make a difference.

Our Diverse Backgrounds: We, the authors of this book, come from different parts of the globe, bringing a wealth of intellectual, financial, social, and political experiences to the table. Each of our stories is unique yet intertwined with a common goal: to leverage our diverse backgrounds for the betterment of Sierra Leone.

Growing Up and Early Influences: Growing up in Freetown, we were constantly inspired by the resilience of our community. The streets of Freetown and rural areas, often bustling with life and energy, were also marked by profound poverty and uncertainty. Our parents, though not wealthy, instilled in us the value of education and hard work.

Building A Nation

They often spoke about the importance of giving back to our community, and these lessons stayed with us as we pursued higher education in various countries such as the United States, the United Kingdom, Holland, Canada, and Spain, to name just a few.

Our childhoods were filled with challenges that are all too familiar to many Sierra Leoneans. We grew up in neighborhoods where access to clean water was a daily struggle, and electricity was a rare luxury. Despite these hardships, our parents were unwavering in their belief in the power of education. They sacrificed much to ensure we attended school, often working multiple jobs to pay for our school fees and books.

We remember the long walks to school under the hot sun, the days when we went to bed hungry because there was not enough food, and the nights spent studying by candlelight. These experiences, though difficult, shaped our determination and resilience. They taught us the value of perseverance and the importance of community support.

Our parents often reminded us that education was our ticket to a better future, not just for ourselves, but for our entire community. They encouraged us to dream big, to aspire for more, and to always remember where we came from. This sense of duty to our community was a guiding force as we pursued higher education abroad.

Leaving Freetown for the opportunities of higher education in the United States, the United Kingdom, Holland, Canada, and Spain was a daunting journey. We were excited but also apprehensive, stepping into unfamiliar worlds far removed from the streets where we grew up.

Each of us faced unique challenges—navigating new cultures, dealing with homesickness, and managing the pressures of academic life. Yet, the values instilled by our parents kept us grounded. We remained focused on our goals, knowing that our success would pave the way for others back home.

In the bustling cities of New York, Houston, Fargo, London, Amsterdam, Toronto, and Madrid, we found our footing. The rigorous academic environments honed our skills and expanded our horizons. We met people from diverse backgrounds, each with their own stories of struggle and triumph. These interactions enriched our understanding of the world and reinforced our commitment to using our education for the greater good.

Despite the physical distance from Sierra Leone, our hearts remained close to home. We stayed connected with our communities through letters, phone calls, and visits. We celebrated their victories and mourned their losses, all the while striving to acquire the knowledge and skills that would enable us to make a meaningful impact upon our return.

Our experiences abroad were transformative. We learned not only from our professors and textbooks but also from the resilience and determination of fellow international students. We discovered that our struggles were not unique; they were part of a broader narrative shared by many from developing nations. This realization fueled our desire to contribute to the development of Sierra Leone to bring about the change that our parents had always dreamed of.

In summary, our early influences and the journey of pursuing higher education abroad were marked by significant challenges and profound lessons.

The poverty and uncertainty of our childhoods instilled in us a deep sense of purpose and responsibility. Our parents' unwavering belief in the power of education and community support guided us through the toughest times. Today, as we stand on the cusp of making a difference, we carry with us the values and resilience that Freetown taught us, ready to give back and uplift our beloved Sierra Leone.

Facing Common Challenges: Upon moving abroad, we faced the typical challenges of an immigrant: cultural adaptation, financial struggles, and the pressure to succeed. Stepping into new and unfamiliar environments, we had to quickly learn and adapt to different ways of life. The bustling streets of New York, the historic lanes of London, the canals of Amsterdam, the skyscrapers of Toronto, and the vibrant plazas of Madrid—all offered new experiences, but also significant challenges.

Cultural adaptation was one of the first hurdles. We found ourselves in societies with different customs, languages, and social norms. Navigating daily interactions, understanding new educational systems, and fitting into social circles were daunting tasks. The sense of being an outsider was palpable, and we often struggled to find our place. Yet, these challenges also taught us resilience and the ability to navigate diverse environments.

Financial struggles were another significant challenge. Many of us came from modest backgrounds, and studying abroad required substantial financial sacrifices. We took on part-time jobs, scholarships, and loans to fund our education. Balancing work and study was often overwhelming, but it was a necessary step to achieve our dreams. The pressure to succeed academically while managing financial responsibilities added to the stress.

Like many immigrants, we sent money back home regularly. This was not just an obligation but a heartfelt commitment to support our families and contribute to local projects in Sierra Leone. Remittances from the diaspora are a critical lifeline for many families, covering essential needs like education, healthcare, and housing. However, despite our best efforts, we often felt disconnected from the impact of our contributions.

There were times when we questioned whether our financial support was making a real difference. Decisions about how the funds were used were often made without our input, and we had no way to ensure that our contributions were leading to sustainable development. This sense of disconnection was frustrating. We longed to see tangible results, to know that our hard-earned money was driving positive change back home.

For example, we contributed to community projects, hoping to improve infrastructure or support education initiatives. Yet, without proper channels of communication and transparency, we were left in the dark about the outcomes of these projects. This lack of feedback made it difficult to gauge the effectiveness of our efforts. We wanted to do more than just send money; we wanted to be actively involved in shaping the future of our communities.

These experiences of cultural adaptation, financial struggles, and the pressure to succeed are shared by many in the diaspora. The journey was challenging, but it also reinforced our determination to make a difference. We learned valuable lessons about perseverance, resilience, and the importance of staying connected to our roots.

Moving forward, we realized the need for better structures to engage the diaspora. Platforms that facilitate clear communication, transparency, and active involvement can bridge the gap between aspiration and action. By sharing our experiences and working collectively, we can ensure that our contributions lead to meaningful and lasting change in Sierra Leone.

In summary, the challenges we faced upon moving abroad were significant but also formative. Cultural adaptation, financial struggles, and the pressure to succeed shaped our resilience and deepened our commitment to supporting our communities. Despite the sense of disconnection at times, these experiences have driven us to seek better ways to engage with and contribute to the development of Sierra Leone.

Finding Strength and Clarity: During these challenges, we found strength in community engagement. Amidst the cultural adaptation, financial struggles, and the pressures of academic life, we sought solace and support in connecting with others who shared our experiences and aspirations. Community engagement became a lifeline, providing us with the clarity and resilience needed to navigate our journey.

We attended workshops and conferences focused on diaspora contributions and development. These events were more than just educational sessions; they were platforms for building connections and fostering a sense of collective purpose. Through these gatherings, we broadened our understanding of the potential impact the diaspora could have on Sierra Leone's development.

One significant conference was the African Diaspora Investment Symposium, where experts and leaders discussed the vital role of the diaspora in Africa's growth. The symposium emphasized the importance of leveraging diaspora resources, skills, and investments to drive sustainable development. Attending such events opened our eyes to the myriad ways we could contribute beyond just financial remittances.

Through these experiences, we connected with like-minded individuals who were equally passionate about making a difference in Sierra Leone. We met fellow Sierra Leoneans and other Africans who shared our vision of a prosperous and stable homeland. These interactions were incredibly inspiring and motivating. We exchanged ideas, shared experiences, and developed strategies for collective action.

One memorable workshop was organized by the Sierra Leone Diaspora Network. It focused on building effective community organizations and advocacy groups to influence policy and drive development projects. The workshop provided practical tools and insights on how to mobilize resources, engage with policymakers, and implement sustainable initiatives. These sessions were invaluable, equipping us with the knowledge and skills needed to make a tangible impact.

Engaging with the diaspora community also helped us find clarity in our individual and collective goals. It was reassuring to see that we were not alone in our struggles and aspirations. The shared sense of purpose and determination reinforced our belief in the power of collective action. We realized that, together, we could overcome challenges and drive significant change in Sierra Leone.

We also participated in local community activities, joining Sierra Leonean associations in our respective cities. These groups organized cultural events, fundraisers, and advocacy campaigns, fostering a sense of unity and shared identity. Being part of these communities helped us stay connected to our roots and provided a support network that was crucial during tough times.

In summary, community engagement gave us the strength and clarity needed to navigate our journey. Through workshops, conferences, and local community activities, we broadened our understanding of diaspora contributions and connected with like-minded individuals passionate about making a difference in Sierra Leone. These experiences reinforced our commitment to collective action and equipped us with the tools to drive sustainable development.

Design of a sustainable Development friendly Education Plan

The greatest challenge Sierra Leone facing is the availability of knowledge that is adaptable to the cultural and social needs, and which is development friendly. It shall be therefore a priority of the diaspora to devise models and lobbying for instituting a sustainable culturally adaptable type of education that addresses the needs of all and ensures every talent is properly catered for and developed to instigate self-reliance and a trigger of the private sector which in every country runs the economy.

The model shall put premium on type of and diversification of curriculum, equipment and provision of teaching materials, training and sustenance of quality teaching and as an addenda introduction of an apprenticeship and internship program to provide the young with adequate work ethics before they enter the job market. This model shall also provide doable funding approaches that can sustain the cost of education over the years.

Advocacy for an effective Agricultural and Rural development Plan that is Sustainable and Scalable.

Food security and the effective distribution of the population to ensure a trigger of the private agro-industrial sector is a key area of development that needs premium attention. The diaspora and home based shall collaborate on the design of and lobby for implementation of a model that shall ensure the following:

- Agriculture and agro industrialisation get the priority they deserve to assure food security, economic growth and development of rural communities.

- Initiate infrastructural and economic projects that shall initiate an urban to rural internal migration for the 65% young and able bodied to migrate to areas they can be gainful employed or engaged in activities of self-development. This shall be based on a coordinated plan that includes training, apprenticeship and even the National Youths Service which shall be a training ground for building a national cultural and technological consciousness and patriotic attitude of the young and future generations.

- Draw up a sustainable and scalable plan for rural development that provides the basis amenities and infrastructure for people to be comfortable living in rural areas whilst they enjoy ease of economic and other activities that promote their ability to participate in wealth creation, distribution and sustenance of equity for growth and development.

Organizing Fundraisers for Healthcare Projects: One of the key initiatives we undertook was organizing fundraisers for healthcare projects in Sierra Leone. The healthcare system in our homeland faces numerous challenges, including a lack of medical supplies, inadequate facilities, and insufficient healthcare professionals. Recognizing these issues, we decided to act by raising funds to support healthcare initiatives.

We began by organizing community events such as charity walks, cultural festivals, and online crowdfunding campaigns. These events brought together members of the diaspora and local communities, fostering a sense of unity and purpose. Through these fundraisers, we were able to collect significant amounts of money, which were then used to purchase medical supplies and fund essential healthcare projects back home.

One notable success was our partnership with a local NGO that focuses on maternal and child health. By channeling the funds raised, we helped establish a maternity clinic in a rural area, providing much-needed services to hundreds of women and children. This project not only improved healthcare access but also demonstrated the power of diaspora engagement in addressing critical needs.

Participating in Policy Advocacy Groups: In addition to fundraising, we actively participated in policy advocacy groups to influence positive change in Sierra Leone. These groups were platforms where we could voice our concerns, share our expertise, and advocate for policies that promote sustainable development.

One such group was the African Diaspora Policy Centre, which focuses on empowering diaspora communities to influence policies in their home countries. Through our involvement, we learned how to effectively engage with policymakers, draft policy recommendations, and mobilize community support. This experience was instrumental in enhancing our advocacy skills and understanding the policy-making process.

We also worked with organizations like the Sierra Leone Policy Watch, which provides a forum for discussing and addressing key policy issues affecting Sierra Leone. By contributing to policy discussions and collaborating with experts, we gained valuable insights into the challenges and opportunities in various sectors, including education, healthcare, and economic development.

Coaching Sessions and Mentorship: Coaching sessions and mentorship played a significant role in our journey. We sought guidance from seasoned professionals who had extensive experience in international development and diaspora engagement. These mentors provided us with practical advice, shared their experiences, and helped us navigate the complexities of our initiatives.

Through coaching, we learned how to design and implement effective development projects, manage resources efficiently, and measure the impact of our efforts.

For instance, we attended workshops on project management and grant writing, which were crucial in securing funding and ensuring the successful execution of our projects.

One mentor, a former UN development officer, guided us through the intricacies of working with international organizations and donors. This mentorship was invaluable in building our capacity to engage with global development networks and leverage their resources for local projects.

The Impact of a Coordinated Effort: A real-life example of our efforts is the establishment of a vocational training center in Freetown and interior region of Sierra Leone. This project was born out of our desire to address the high unemployment rate among young people in Sierra Leone. By providing vocational training in areas such as carpentry, tailoring, and information technology, the center aims to equip young people with the skills needed to secure employment and improve their livelihoods.

The project was a collaborative effort involving fundraisers, advocacy, and mentorship. We organized fundraising events to gather the necessary funds, advocated for policy support from local authorities, and received guidance from development experts on setting up and managing the center. Today, the vocational training center is thriving, providing opportunities for hundreds of young Sierra Leoneans and contributing to the local economy.

Conclusion: In summary, our involvement in various initiatives, from organizing fundraisers to participating in policy advocacy groups, provided us with practical insights and strategies that were crucial in our journey.

Through coaching sessions and mentorship, we gained the expertise needed to navigate the complexities of international development and effectively channel resources for maximum impact. These experiences not only enriched our personal and professional lives but also empowered us to make a meaningful difference in Sierra Leone.

Helping Others Along the Way: Over the years, we've had the privilege of helping other Sierra Leoneans in the diaspora overcome similar challenges. Our journey has been marked by numerous instances of providing support, guidance, and strategic advice to fellow Sierra Leoneans striving to make a difference. Whether it was offering insights on starting a nonprofit, assisting with community projects, or simply providing a listening ear, these experiences have reinforced our belief in the power of collective action. Even when we weren't paid for our efforts, the satisfaction of seeing tangible results kept us motivated.

Providing Guidance on Starting Poverty Reduction Projects: One of the most rewarding aspects of our journey has been helping others start their own social and economic growth and development. Many Sierra Leoneans in the diaspora are passionate about contributing to the development of their homeland but often lack the knowledge or resources to get started. We have guided numerous individuals through the process of establishing through nonprofit and for-profit initiatives by developing researched models and theories of change that are designed to impact the poor, are sustainable and scalable..

For instance, we assisted a group of young Sierra Leoneans in the UK in setting up a nonprofit aimed at providing educational resources to underprivileged schools in Sierra Leone. We helped them develop a clear mission, create a strategic plan, and organize fundraising events. Today, their organization has successfully delivered books, computers, and other educational materials to several schools, significantly improving the learning environment for many children.

Offering Strategic Advice for Community Projects: In addition to helping start nonprofits, we have also provided strategic advice for various community projects. These projects range from healthcare initiatives to agricultural programs, each aimed at addressing specific needs within Sierra Leone. By sharing our experiences and insights, we have enabled others to design and implement effective projects that make a real difference.

One memorable project involved supporting a group in the United States that wanted to set up a clean water initiative in a rural area of Sierra Leone. We advised them on project management, fundraising strategies, and ways to engage local communities. With our guidance, they were able to drill several wells, providing clean drinking water to thousands of people and reducing the incidence of waterborne diseases.

Mentoring and Networking: Mentoring has been another crucial aspect of our efforts to help others. Through one-on-one sessions and group workshops, we have mentored aspiring leaders and changemakers, sharing our knowledge and experiences to help them succeed. These mentoring relationships often extended beyond formal sessions, developing into long-term partnerships and friendships.

We have also leveraged our networks to connect individuals with resources and opportunities. By introducing them to potential donors, partners, and advisors, we have helped many diaspora members amplify their impact. For example, we connected a Sierra Leonean entrepreneur in Canada with investors who funded her innovative agricultural project, which now supports hundreds of farmers in Sierra Leone.

Empowering a Community Health Initiative: One particularly impactful experience was supporting a community health initiative led by a group of Sierra Leonean doctors in Germany. They wanted to address the high maternal and child mortality rates in rural Sierra Leone but needed guidance on structuring and funding their project. We worked closely with them, providing strategic advice on project design, grant writing, and stakeholder engagement.

Through our support, they secured funding from international health organizations and successfully launched their initiative. Today, their project operates several health clinics, offering vital maternal and child health services to communities that previously had little or no access to healthcare. The tangible improvements in health outcomes and the empowerment of local healthcare workers have been immensely rewarding to witness.

The Power of Collective Action: These experiences have reinforced our belief in the power of collective action. By helping others, we have not only contributed to the development of Sierra Leone but also built a strong network of engaged and empowered individuals committed to making a difference. The success of these initiatives demonstrates that when we work together, we can achieve far more than we could alone.

The satisfaction of seeing tangible results from our collective efforts keeps us motivated. Each successful project, each improved community, and each empowered individual serves as a reminder of the impact we can have when we come together for a common cause. Our journey has shown that by sharing our knowledge, resources, and support, we can create lasting change and inspire others to do the same.

In summary, helping others along the way has been a cornerstone of our journey. Through guidance, strategic advice, mentoring, and networking, we have empowered fellow Sierra Leoneans to overcome challenges and drive positive change. These experiences have reinforced our belief in collective action and the immense potential of the diaspora to contribute to Sierra Leone's development. The satisfaction of seeing tangible results from our efforts continues to inspire and motivate us to keep pushing forward.

Looking Ahead: As we look to the future, our goal is to create a structured framework for diaspora engagement. By sharing our stories and strategies in this book, we aim to empower you to act. Imagine a network of Sierra Leoneans across the globe, united in purpose, driving sustainable development through coordinated efforts. This vision is within reach, and together, we can turn it into reality.

Motivational Ending: Remember, your journey may be challenging, but it is also incredibly rewarding. Each step you take brings you closer to making a meaningful impact. The path to change is seldom easy, but it is through perseverance and collective effort that we make the greatest strides. As Nelson Mandela wisely said, "It always seems impossible until it's done."

This powerful statement resonates with our experiences and serves as a reminder that no matter how insurmountable the obstacles may seem, they can be overcome with determination and unity.

Your efforts, no matter how small, are part of a larger movement towards a better Sierra Leone. Every action you take contributes to a ripple effect that can transform communities and uplift the nation. The struggles and challenges you face are shared by many, and together, we can turn these challenges into opportunities for growth and development.

Stay committed, stay inspired, and know that your contributions are invaluable. The journey you embark on is one of resilience, hope, and impact. It is a journey that not only transforms the lives of those in Sierra Leone but also enriches your own life through the connections and achievements you make along the way.

In the next chapter, we will delve into practical steps and strategies that have proven effective in fostering development. We will outline how you can harness your potential and contribute to the transformation of Sierra Leone. By providing concrete examples and actionable plans, we aim to empower you to take meaningful steps towards positive change. Whether it's through community projects, advocacy, or personal initiatives, there are numerous ways you can make a difference.

Join us as we continue this journey, turning aspirations into actionable plans and dreams into reality. Together, we can build a brighter future for Sierra Leone, one step at a time. Your dedication and passion are the driving forces behind this transformation, and with your involvement, we can achieve great things.

Building A Nation

Let's embark on this journey with hope and determination, knowing that our collective efforts will pave the way for a prosperous and just Sierra Leone.

Reflection and Engagement Questions

In this chapter, we delve into the personal journeys of Sierra Leoneans in the diaspora, exploring the challenges and triumphs that have shaped their commitment to their homeland. Through a series of real-life examples and shared experiences, we highlight the resilience and determination that drives many to contribute to Sierra Leone's development. This chapter aims to inspire readers by showcasing the power of collective action and the significant impact that individual efforts can have when supported by a strong, engaged community. It underscores the importance of community engagement, mentorship, and strategic partnerships in overcoming obstacles and making a tangible difference.

1. **Personal Journey:**
    - How has your personal journey and background shaped your views on contributing to Sierra Leone's development?
    - _____
      _____

2. **Facing Challenges:**
    - Reflect on the challenges you've faced living locally or abroad.

How have these experiences influenced your desire to engage with and support Sierra Leone?

   o _____
      _____

3. **Community Support:**

   o In what ways has community engagement provided you with strength and clarity in your efforts to contribute to Sierra Leone?

   o _____
      _____

4. **Learning from Others:**

   o Consider the examples of other Sierra Leoneans or diaspora communities who have successfully initiated projects or advocacy efforts. What lessons can you draw from their experiences to apply to your own initiatives?

   o _____
      _____

5. **Future Contributions:**

   o What specific actions or projects can you envision undertaking to help bridge the gap between the diaspora and Sierra Leone, based on the strategies and insights shared in this chapter?

   o _____
      _____

CHAPTER 3:
ACTIONABLE BLUEPRINT FOR GOVERNANCE REFORM AND DEVELOPMENT

Introduction to the Actionable Blueprint: Although this chapter might be one of the shortest, it is often a deciding chapter for most readers who need our expertise. This chapter gives our ideal reader an idea of what we would be discussing in each step. Instead of thinking of these sections as chapters, we want readers to envision them as steps in a process—an actionable roadmap they can follow to effect change.

Step 1: Understanding Good Governance

Overview: Good governance is the bedrock of any functioning democracy. In this step, we will explore the essential principles of good governance, such as accountability, transparency, responsiveness, and the rule of law. Understanding these principles is crucial for creating a framework where citizens can thrive and trust in their government. By adhering to these core tenets, governments can foster an environment of trust, stability, and progress.

Key Points:

- **Accountability:** Ensuring that government officials are answerable for their actions is fundamental to good governance. Accountability mechanisms, such as regular audits, independent oversight bodies, and clear reporting requirements, help maintain integrity and trust in public institutions. It is essential that public officials are held to high ethical standards and that there are consequences for misconduct. Citizens must be empowered to demand accountability through civic education and participation in governance processes.

- **Transparency:** Making government operations visible to the public is crucial for fostering trust and preventing corruption. Transparency involves the open sharing of information about government activities, decisions, and expenditures. This can be achieved through initiatives such as open data portals, public access to government documents, and transparent procurement processes. When citizens have access to information, they can make informed decisions and hold their leaders accountable.

- **Responsiveness:** Adapting policies and services to meet the needs of the people is a key aspect of responsive governance. Governments must be attuned to the concerns and needs of their citizens, actively seeking feedback and adjusting policies accordingly. This can be facilitated through regular public consultations, surveys, and town hall meetings. Responsive governance ensures that services are delivered efficiently and effectively, enhancing public satisfaction and trust.

- **Rule of Law:** Applying laws impartially to maintain order and protect rights is a cornerstone of good governance.

The rule of law ensures that everyone, including government officials, is subject to the law. It guarantees that laws are enforced fairly and consistently, protecting the rights and freedoms of all citizens. A strong legal framework, independent judiciary, and effective law enforcement agencies are essential components of upholding the rule of law.

- **Maintaining Ethical Standards:** A key component of any good governance system is the ability of public and private sectors to institute and sustain ethical standards for which it is necessary to have a structured Guideline documented. SLAM shall research and develop a document that can be standardized within the cultural needs of Sierra Leone.

By embracing these principles, governments can create a stable and just society where citizens feel secure and empowered. Good governance not only promotes development and prosperity but also enhances the legitimacy and effectiveness of public institutions. In the following steps, we will delve into practical strategies for implementing these principles and building a robust governance framework.

Step 2: Enhancing Governance Through Key Strategies

Overview: Effective governance requires practical strategies that can be implemented to improve existing systems. This step outlines specific strategies such as public monitoring initiatives, advocacy for robust legal frameworks, and the use of technology to increase transparency and efficiency.

By adopting these strategies, governments can enhance their

performance, build public trust, and ensure that governance processes are inclusive and effective.

Key Points:

- **Public Monitoring Initiatives:** Engaging citizens in tracking government performance is crucial for promoting accountability and transparency. Public monitoring initiatives involve citizens in the oversight of government activities, from budget expenditures to service delivery. Tools such as citizen report cards, community scorecards, and social audits can provide valuable feedback on government performance. These initiatives empower citizens to hold their leaders accountable and ensure that public resources are used effectively. For instance, community-driven monitoring of public projects can prevent corruption and ensure that projects are completed to the required standards.

- **Advocacy for Legal Frameworks:** Strengthening laws to support good governance is essential for creating a stable and fair political environment. Advocacy efforts can focus on enacting or reforming legislation to enhance transparency, accountability, and citizen participation. This includes laws that mandate open government practices, protect whistleblowers, and establish independent anti-corruption bodies. Advocacy groups can work with lawmakers, civil society organizations, and international partners to push for legal reforms that support good governance. Effective legal frameworks provide the foundation for sustainable governance improvements.

- **Educational Programs:** Informing citizens about their rights and responsibilities is a fundamental aspect of fostering an informed and active citizenry. Educational programs can take various forms, including workshops, seminars, school curricula, and public awareness campaigns. These programs should aim to increase citizens' understanding of governance processes, their rights under the law, and the mechanisms available for them to participate in governance. Educated citizens are better equipped to advocate for their rights, engage in public discourse, and hold their leaders accountable.

- **Technology Platforms:** Using digital tools to enhance transparency and engagement is a powerful strategy for modern governance. Technology platforms such as government websites, mobile apps, and social media can provide citizens with easy access to information and services. E-governance initiatives can streamline administrative processes, reduce corruption, and improve service delivery. For example, online portals for public procurement can ensure that bidding processes are open and transparent. Additionally, social media platforms can facilitate real-time communication between government officials and the public, fostering greater engagement and responsiveness.

By implementing these strategies, governments can significantly improve their governance systems, making them more transparent, accountable, and effective. Public monitoring initiatives, robust legal frameworks, educational programs, and technology platforms are practical tools that can drive meaningful change.

In the next steps, we will explore how these strategies can be tailored and applied to specific governance challenges, ensuring that the principles of good governance are upheld and that citizens' needs are met.

Step 3: Building Democratic Values and Social Democratization

Overview: To ensure the sustainability of democratic values, it is crucial to introduce a mass social education for society to adopt and sustain democratic principles and perceptions. First it must be understood that democracy is not political, but a social belief and perception reflected in the attitudes of the people. Second it is important to understand that democracy is the application of inclusive empathy and tolerance so that every member of society can use their opinion for, to and by all for a common good. Thus, each must be able to make choices for all, to all and so be responsible to be by the general. When and where such social attitudes are used in making political decisions then collective participation is enhanced thereby sustaining a society of trust and chain processes that allow each and all to thrive and prosper. This step focuses on maintaining long-term goals, fostering inclusivity, ensuring accountability, and building consensus. Upholding these principles helps to create a resilient and adaptive governance structure that can respond effectively to the needs and aspirations of all citizens.

Key Points:

- **Long-term Goals:** Focusing on sustainable development rather than short-term gains is essential for the prosperity of a nation. Building mass democratic thinking and perceptions to enhance good governance should prioritize in policies and initiatives that promote long-term social, participation, sustainable, sustainable economic growth and development, and environmental management are a prerequisite through a long-term roadmap that should be legally mandatory on governments to continue to implement and complete. This includes investing in the correct type of education that can allow every talent to be developed, healthcare for all, infrastructure and telecommunications, and sustainable environmental conservation. By setting long-term goals, governments can ensure that their actions today lay the foundation for a better future. Planning for the long term helps to stabilize economies, improve quality of life, and protect resources for future generations.

- **Inclusivity:** Ensuring that all segments of society are represented and heard is a cornerstone of democracy. Inclusivity means actively engaging marginalized groups, including women, youth, ethnic minorities, and people with disabilities, in the decision-making process. It involves creating platforms and mechanisms for diverse voices to be heard and considered in policy formulation. Inclusivity fosters social cohesion and ensures that policies are reflective of the needs and aspirations of the entire population.

Governments can implement affirmative action policies, support civil society organizations, and promote inclusive dialogues to enhance participation.

- **Accountability and Transparency:** Maintaining clear and open governance processes is vital for building trust and legitimacy. Accountability ensures that government officials are held responsible for their actions, while transparency ensures that government operations are open to public scrutiny. Together, these principles help to prevent corruption, enhance efficiency, and promote ethical behavior in public service. Measures to enhance accountability and transparency include public disclosure of government spending, regular audits, independent oversight bodies, and access to information laws. When citizens can see and understand how decisions are made and resources are allocated, they are more likely to trust and support their government.

- **Building Consensus:** Mediating diverse interests to reach common ground is essential for effective governance. In a democratic society, differing opinions and interests are inevitable. Building consensus involves negotiating and mediating these differences to arrive at decisions that are acceptable to all parties. This process requires strong leadership, effective communication, and a commitment to dialogue. Consensus-building helps to prevent conflicts and ensures that policies are more robust and sustainable. Techniques for building consensus include inclusive, deliberative processes, participatory decision-making, and conflict-resolution mechanisms.

By upholding these democratic principles, governments can create a governance structure that is resilient and adaptive. This structure not only responds to current challenges but also anticipates and prepares for future needs. Maintaining long-term goals, fostering inclusivity, ensuring accountability, and building consensus are essential steps towards a stable, just, and prosperous society. In the next steps, we will discuss practical strategies for embedding these principles into governance practices ensuring that ensure democracy is not only sustained but also thrives.

Step 4: Combating Detrimental Principles

Overview: Identifying and addressing harmful governance practices is essential for creating a healthy political environment. This step discusses strategies to combat corruption, nepotism, political partisanship, and other detrimental principles that hinder progress. By addressing these issues, we can create a more equitable, efficient, and trustworthy government that truly serves the interests of all citizens.

Key Points:

- **Combating Corruption:** Implementing anti-corruption measures and promoting ethical standards is critical for good governance. Corruption erodes public trust, diverts resources from essential services, and undermines the rule of law. Effective anti-corruption strategies include establishing independent anti-corruption bodies, enforcing strict penalties for corrupt practices, and promoting a culture of integrity within public institutions.

Transparency International and similar organizations provide frameworks and tools that can help governments combat corruption. Additionally, public education campaigns can raise awareness about the negative impacts of corruption and encourage citizens to report corrupt activities.

- **Addressing Nepotism:** Ensuring merit-based appointments and opportunities is vital to create a fair and competent public service. Nepotism undermines the effectiveness of government institutions by prioritizing personal connections over qualifications and merit. To address this, governments should implement transparent recruitment processes, establish clear criteria for appointments, and promote a culture of meritocracy. This can include publicizing job openings, conducting impartial selection procedures, and providing training programs to develop the skills of all employees. By fostering a merit-based system, governments can ensure that the most qualified individuals are in positions of power, leading to more efficient and effective governance.

- **Reducing Partisanship:** Encouraging collaboration across political lines is essential for creating a cohesive and functional government. Political partisanship can lead to gridlock, hinder policy implementation, and exacerbate divisions within society. To reduce partisanship, governments should promote bipartisan or multi-party dialogue, create coalition governments where necessary, and encourage joint problem-solving initiatives. Engaging in regular consultations with all political parties and fostering a culture of respect and cooperation can help bridge divides.

By prioritizing the common good over partisan interests, governments can develop more comprehensive and widely supported policies.

- **Promoting Transparency:** Making government actions visible to prevent misconduct is a fundamental principle of good governance. Transparency ensures that government operations are open to public scrutiny, which helps deter unethical behavior and build public trust. Strategies to promote transparency include adopting open data initiatives, ensuring public access to government records, and requiring detailed disclosure of government spending and decision-making processes. Additionally, establishing independent media and civil society organizations that can monitor and report on government activities is crucial. Transparency measures should be supported by laws that protect whistleblowers and guarantee the right to information.

By effectively combating detrimental principles such as corruption, nepotism and partisanship and by promoting transparency, governments can create a healthier political environment. These measures not only improve the efficiency and fairness of governance but also enhance public trust and engagement. In the subsequent steps, we will explore how to implement these strategies in practice, providing actionable insights and examples of successful initiatives. Through these efforts, we can build a government that is accountable, inclusive, and dedicated to serving the people.

Step 5: Mobilizing the Diaspora for Development

Overview: The Sierra Leonean diaspora holds significant potential for national development. This step outlines how the diaspora, in collaboration with local and international partners, can be mobilized to contribute to the country's progress through investments, advocacy, and skills transfer. By leveraging the strengths and resources of the diaspora alongside those of local and international entities, we can create a synergistic impact that drives sustainable development.

Key Points:

- **Investment Opportunities:** Encouraging diaspora investments in local projects is crucial. By identifying and promoting viable investment opportunities, we can attract funds that support economic growth and infrastructure development. Collaborations with local businesses and international investors can enhance these efforts, ensuring that investments are well-placed and impactful.

- **Advocacy and Influence:** Leveraging the diaspora's voice in international forums is vital for raising awareness and advocating for Sierra Leone's needs. Diaspora members can act as ambassadors, influencing policy decisions and garnering support from international organizations. By forming alliances with local advocacy groups and international NGOs, the diaspora can amplify its voice and impact.

- **Skills Transfer:** Facilitating the exchange of knowledge and expertise between the diaspora and local professionals is key to building capacity.

Programs that enable diaspora experts to share their skills through workshops, mentoring, and collaborative projects can significantly enhance local capabilities. Partnering with educational institutions and international development agencies can help formalize and expand these initiatives.

- **Collaborative Initiatives:** Building partnerships between diaspora groups, local entities, and international partners is essential for sustainable development. Joint initiatives that address critical issues such as healthcare, education, and economic development can leverage the strengths of each group. Establishing networks and platforms for collaboration can streamline efforts and maximize the collective impact.

By mobilizing the diaspora in partnership with local and international stakeholders, we can drive meaningful and lasting development in Sierra Leone. This collaborative approach ensures that all available resources and expertise are harnessed to create a brighter future for the nation.

Step 6: Practical Steps for Immediate Action

Overview: In this step, we provide practical, actionable steps that readers can take to begin making a difference immediately. From community organizing to engaging in policy advocacy, these steps are designed to empower readers to take direct action. By equipping individuals with the tools and strategies needed to effect change, we can collectively drive progress and improve governance at the local, national, and international levels.

Key Points:

- **Community Organizing:** Building local networks to address community issues is a foundational step towards creating impactful change. Community organizing involves rallying individuals around common causes, identifying local challenges, and developing collective solutions. This can be achieved by forming neighborhood associations, community groups, and task forces that focus on specific issues such as healthcare, education, or infrastructure. Effective community organizing requires strong leadership, clear communication, and active participation from community members. By fostering a sense of ownership and collective responsibility, communities can address their problems more effectively and sustainably.

- **Policy Advocacy:** Influencing policy changes through organized efforts is a powerful way to drive systemic change. Policy advocacy involves raising awareness about key issues, lobbying policymakers, and pushing for legislative reforms. This can be done through various methods such as writing policy briefs, organizing advocacy campaigns, engaging with media, and building coalitions with like-minded organizations. Advocacy efforts should be well-researched, strategic, and focused on achievable goals. By presenting compelling evidence and mobilizing public support, advocates can persuade decision-makers to adopt policies that promote good governance and social justice.

- **Educational Initiatives:** Promoting awareness and education on key issues is crucial for empowering citizens and fostering an informed electorate. Educational initiatives can take many forms, including workshops, seminars, public lectures, and online courses. These programs should aim to increase understanding of governance processes, citizen rights, and the importance of civic engagement. Collaborating with schools, universities, and civil society organizations can help expand the reach and impact of these initiatives. Educated citizens are better equipped to participate in governance, advocate for their rights, and hold their leaders accountable.

- **Sustainable Projects:** Implementing projects that provide long-term benefits is essential for creating lasting change. Sustainable projects address immediate needs while also building the foundation for future development. Examples include renewable energy installations, community health centers, educational facilities, and agricultural programs. These projects should be designed with sustainability in mind, ensuring that they are environmentally friendly, economically viable, and socially inclusive. Engaging local communities in the planning and execution of these projects can enhance their relevance and effectiveness. Additionally, securing funding from diverse sources, including government grants, private investments, and international aid, can ensure the longevity and scalability of these initiatives.

By taking these practical steps, individuals and communities can begin to make a tangible impact on governance and development.

Community organizing, policy advocacy, educational initiatives, and sustainable projects are powerful tools for driving change and improving the quality of life for all citizens. In the following steps, we will delve deeper into each of these strategies, providing detailed guidance and real-world examples to help you implement them effectively. Together, we can build a more just, inclusive, and prosperous society.

Conclusion and Next Steps: As we move forward, it's essential to understand that transformation begins with each one of us. Everyone has the power to effect change, and collective action amplifies this potential. By examining real-life examples and proven strategies, we can draw valuable lessons on how to effectively contribute to national development. Our journey, along with those of many others, serves as a testament to the power of determination and collective effort.

Real-Life Examples and Proven Strategies: Throughout this book, we have highlighted numerous examples of successful initiatives and effective strategies that have driven meaningful change. From grassroots community organizing to high-level policy advocacy, these stories illustrate the impact of well-coordinated efforts and the importance of perseverance. By learning from these examples, we can identify best practices and adapt them to our specific contexts.

Practical Insights and Actionable Steps: The upcoming steps aim to equip you with the tools necessary to make a significant impact. We will delve into practical insights and actionable steps that you can take to contribute to the development of Sierra Leone. These steps are designed to be clear, achievable, and impactful, providing a roadmap for individuals and communities to follow.

Whether you are a member of the diaspora, a local activist, or an international partner, there are specific actions you can take to drive progress.

Turning Aspirations into Actionable Plans: Our aspirations for a better Sierra Leone must be translated into concrete plans and actions. This requires careful planning, strategic thinking, and a commitment to follow through. By setting clear goals, developing detailed plans, and mobilizing resources, we can turn our dreams into reality. Each step we take brings us closer to achieving our vision of a prosperous and just Sierra Leone.

Building a Brighter Future Together: Together, we can build a brighter future for Sierra Leone, one step at a time. Your dedication and passion are the driving forces behind this transformation. By working collectively, we can overcome challenges, leverage opportunities, and create lasting change. Our collective efforts will pave the way for a nation where all citizens can thrive and enjoy the fruits of development.

A Call to Action: Let's embark on this journey with hope and determination, knowing that our collective efforts will pave the way for a prosperous and just Sierra Leone. Each of us has a role to play, and by taking action, we can make a difference. Join us as we continue this journey, turning aspirations into actionable plans and dreams into reality. Your involvement is crucial, and together, we can achieve great things.

In conclusion, this book provides a comprehensive roadmap for contributing to Sierra Leone's development. By following the steps outlined, you can be part of the solution, driving progress and fostering positive change.

Let's commit to this journey, support one another, and work tirelessly towards building the Sierra Leone we envision. With determination, collaboration, and a shared vision, we can create a brighter future for all.

Reflection and Engagement Questions

This chapter provides a detailed roadmap for implementing governance reforms and fostering development in Sierra Leone. It breaks down the steps necessary to achieve good governance, highlighting the importance of accountability, transparency, responsiveness, and the rule of law. The chapter also emphasizes the role of practical strategies, such as public monitoring initiatives and leveraging technology, to enhance governance. It aims to equip readers with the knowledge and tools needed to take immediate, actionable steps towards creating a more inclusive and effective governance framework. By mobilizing the diaspora and fostering collaborative efforts, this chapter envisions a sustainable transformation that addresses both current and future challenges.

1. **Understanding Principles:**

    - How do the principles of good governance, such as accountability and transparency, apply to the current political and social landscape of Sierra Leone?

    - _____
      _____

2. **Practical Strategies:**
   - Which of the practical strategies outlined, such as public monitoring initiatives or technology platforms, do you think would be most effective in enhancing governance in Sierra Leone? Why?

   - _____
     _____

3. **Mobilizing the Diaspora:**
   - In what ways can the Sierra Leonean diaspora be more effectively mobilized to contribute to national development and governance reforms?

   - _____
     _____

4. **Overcoming Challenges:**
   - Reflect on the challenges discussed in implementing governance reforms. What solutions or approaches can be taken to overcome these obstacles and ensure sustainable change?

   - _____
     _____

5. **Personal Involvement:**
   - Considering the actionable steps provided, what specific actions or initiatives can you personally undertake to support governance reform and development in Sierra Leone? How can you collaborate with others to maximize impact?

   - _____
     _____

CHAPTER 4:
ESTABLISHING THE FRAMEWORK

A Brief History of Sierra Leone's Governance: Sierra Leone, a country with a rich history dating back to the 15th century, has experienced various forms of governance, from pre-colonial traditional systems to British colonial rule and finally to an independent democratic state. The journey of governance in Sierra Leone is marked by significant milestones that have shaped its current political and social fabric.

Colonial Era and Independence: During the colonial period, Sierra Leone was governed by the British, who established Freetown as a colony for freed African slaves in 1787. The colonial administration introduced Western-style governance structures, but these often operated alongside traditional African systems. This dual system sometimes led to conflicts and a complex socio-political environment.

Sierra Leone gained independence from British colonial rule on April 27, 1961. The new nation inherited a parliamentary system of government, with Sir Milton Margai becoming the first Prime Minister. The post-independence era initially saw a relatively stable political environment, but this was soon challenged by internal conflicts and power struggles.

Post-Independence Governance and Challenges: The post-independence period in Sierra Leone was marked by political instability, economic difficulties, and social upheaval.

In 1967, a military coup led to a series of changes in government, which culminated in a one-party state in 1978. The concentration of power and lack of political freedoms during this period sowed the seeds for future conflicts.

The most significant challenge to Sierra Leone's governance came in the form of a brutal civil war that lasted from 1991 to 2002. The war, fueled by grievances over corruption, inequity, and the marginalization of certain groups, resulted in devastating human and economic losses. The conflict ended with the help of international intervention and the establishment of a transitional government that aimed to restore peace and democratic governance.

Rebuilding and Democratic Transition: The end of the civil war marked the beginning of a rebuilding phase for Sierra Leone. The establishment of the Truth and Reconciliation Commission (TRC) and the Special Court for Sierra Leone were pivotal in addressing the war's atrocities and promoting national healing. In 2002, the country held its first post-war democratic elections, which were widely seen as free and fair, setting the stage for a new era of democratic governance.

Despite these efforts, Sierra Leone continues to face significant challenges. Corruption remains a pervasive issue, undermining public trust in government institutions. Economic development has been slow, with many citizens still living in poverty. Social issues such as youth unemployment, health crises (like the Ebola outbreak), and gender inequality also pose serious threats to national stability and progress.

Introducing SLAM: The Sierra Leone Advocacy Movement was established to mobilize the Sierra Leonean diaspora and foster a collaborative approach to national development. The organization aims to maximize Sierra Leone's potentials by promoting democratic principles, social justice, and economic development. SLAM believes in an open and democratic society where every citizen can develop and enjoy their talents without fear or undue influence.

Navigating the Complex Landscape of National Unity Governments: Insights and Predictions: The concept of a Government of National Unity (GNU) often emerges as a hopeful solution to political crises marked by division and strife. Advocacy groups such as SLAM are expected to participate and foster peace, stability, and negotiation in such situations. However, the process can quickly become complicated when there is a lack of trust. The practical realization of such governments frequently underscores the adage that forming a true, functioning unity government is easier said than done. This narrative examines various global examples to explore the inherent challenges and predictability of GNUs, raising critical questions about their feasibility and durability in politically fragmented environments.

Defining National Unity: Ambitions and Realities - At its core, a GNU is intended to foster broad-based inclusivity and mitigate conflict by incorporating diverse political factions into a single governing body. This concept is appealing, particularly in post-conflict or deeply divided societies where traditional single-party governments may exacerbate existing tensions. However, the implementation of such governments reveals a complex interplay of ambition versus reality, where the idealistic goals of unity face the pragmatic challenges of political dynamics.

Global Examples: Lessons Learned

1. **Zimbabwe (2009-2013):** Zimbabwe's GNU was established following a contested election, aiming to stabilize the nation and halt economic freefall. Initially, this government, which included rival parties ZANU-PF and MDC, appeared as a beacon of hope. However, the unity facade soon crumbled under the weight of power monopolization by ZANU-PF, highlighting a significant issue with GNUs: the retention of power tends to trump the sharing of it. This example illustrates how GNUs can fail when underlying motives are skewed towards maintaining existing power structures rather than genuinely democratizing power distribution.

2. **South Africa (1994-1999):** Contrasting sharply with Zimbabwe, South Africa's GNU under Nelson Mandela is heralded as a successful model. It navigated the end of apartheid by fostering reconciliation and ensuring political inclusiveness. Mandela's government not only quelled fears of racial conflict but also laid down foundations for a more equitable society. Here, the strength lay in leadership committed to the vision of unity and a robust legal framework that supported equitable power sharing. Yet, even this successful model was not without its challenges, particularly in addressing deep-seated economic disparities and managing expectations in a transitioning society.

3. **Lebanon's Confessionalist System:** Lebanon presents a unique case of a permanent sectarian power-sharing system intended to prevent dominance by any single group.

While successful in preventing large-scale sectarian conflict, this system has also led to political paralysis and entrenched sectarian divisions more deeply, suggesting that such models of government unity might actually solidify divisions rather than resolve them.

4. **Northern Ireland's Power-Sharing Executive:** Northern Ireland's power-sharing model, established by the Good Friday Agreement, has oscillated between cooperation and crisis. The model has successfully mitigated sectarian violence but has been plagued by political stalemates and suspensions of the Assembly, pointing out the delicate balance required to maintain unity in a landscape of entrenched historical divisions.

Predictability and Warning Signs: From these examples, several factors emerge that can predict the success or failure of GNUs:

1. **Leadership Commitment:** Successful GNUs, like South Africa's, are driven by leaders genuinely committed to the ethos of unity.

2. **Institutional Framework:** Robust legal and institutional frameworks that support equitable power sharing and conflict resolution are crucial.

3. **Underlying Motives:** GNUs fail when underlying motives are skewed towards maintaining existing power structures.

4. **Societal Support:** Broad societal support and a willingness among the populace to move past division are essential for sustaining GNUs.

Predicting the outcome of a GNU involves identifying these factors and understanding their interplay within the specific political context of a country. Warning signs often missed include escalating unilateral decisions by dominant parties, lack of transparency in decision-making, and failures in implementing agreed-upon reforms.

As such, the feasibility and effectiveness of Governments of National Unity hinge on a complex set of factors, including leadership quality, institutional support, and genuine commitment to shared governance. As global examples illustrate, while GNUs can offer temporary stability and a pathway out of conflict, their long-term success in fostering true national unity is contingent upon continuous and committed efforts across all sectors of society. Predicting their success involves not just hopeful optimism but a grounded analysis of political motives and societal readiness to embrace such profound changes. This narrative, therefore, serves not only to question the viability of GNUs but also to offer a blueprint for assessing and implementing these complex governance structures in the hope of achieving lasting peace and unity.

Current Challenges and Opportunities: Today, Sierra Leone stands at a crossroads. While significant progress has been made in building democratic institutions and promoting political stability, the nation still grapples with numerous challenges. The need for effective governance, transparent leadership, and inclusive development remains critical.

In this context, the role of the Sierra Leonean diaspora has become increasingly important. Organizations like SLAM are crucial in leveraging the resources, expertise, and networks of the diaspora to support the country's development.

Case Study: Effective Inquiry and Reforms at the Sierra Leone Roads Authority (SLRA)

Introduction: The Sierra Leone Roads Authority (SLRA) is a noteworthy example of how a well-structured inquiry and subsequent reforms can significantly improve governance and operational efficiency. Established to address the challenges of road maintenance and development, the SLRA's journey highlights the importance of strategic reforms, transparency, and stakeholder engagement. This case study provides insights into the success, lessons learned, and recommendations for implementing similar reforms in other government departments.

Establishment and Early Challenges: The SLRA was formed in 1993, following the enactment of the SLRA Act in 1992. This reform separated the Roads Branch from the Ministry of Works, creating a semi-autonomous agency focused solely on road infrastructure. The initial challenges included overstaffing, inefficient resource utilization, and significant government interference, which hindered the agency's ability to operate effectively.

Strategic Reforms and Successes

1. **Road Fund Establishment**: One pivotal reform was the creation of a dedicated road fund. By allocating a portion of the revenue from each gallon of fuel sold to road maintenance, the SLRA secured a steady funding source. This approach significantly improved road maintenance efforts and reduced dependency on unpredictable government allocations.

2. **Major Infrastructure Projects**: The SLRA successfully managed several major road projects, including reconstructing critical routes such as the Kailahun - Koindu roads and other essential highways. These projects enhance connectivity and spur regional economic growth by improving access to markets and services.

3. **Environmental and Social Impact Assessments (ESHIA)**: The SLRA's commitment to conducting comprehensive Environmental and Social Impact Assessments for significant projects ensured that potential adverse impacts were identified and mitigated. This practice demonstrated the agency's dedication to sustainable development and compliance with regulatory standards.

4. **Stakeholder Engagement and Transparency**: The SLRA emphasized the importance of engaging with local communities, government bodies, and international donors. Regular public consultations, transparent reporting, and adherence to environmental guidelines fostered trust and collaboration among stakeholders.

Lessons Learned

1. **Importance of Autonomy**: Granting operational autonomy to agencies like the SLRA can lead to more focused and efficient management. Reduced political interference allows for better strategic planning and execution.

2. **Sustainable Funding Mechanisms**: Establishing dedicated funds, such as the road fund, ensures continuous financial support for critical infrastructure projects. This model can be replicated in other sectors to provide consistent funding for essential services.

3. **Comprehensive Impact Assessments**: Conducting thorough environmental and social impact assessments before initiating projects helps identify potential risks and implement mitigation measures, leading to sustainable development outcomes.

4. **Stakeholder Collaboration**: Engaging with a broad range of stakeholders, including local communities, government officials, and international partners, is crucial for the successful implementation of reforms. Transparency and open communication build trust and foster collaborative efforts.

Recommendations for Future Reforms

1. **Enhanced Autonomy for Agencies**: Future reforms should aim to increase the operational autonomy of government agencies to improve efficiency and reduce political interference.

2. **Expansion of Dedicated Funds**: Similar to the road fund, other sectors should establish dedicated funding mechanisms to ensure steady financial support for their operations and maintenance.

3. **Strengthening Legal Frameworks**: Implementing robust legal frameworks that mandate transparency, accountability, and regular audits will reinforce good governance practices across various departments.

4. **Promoting Best Practices**: Sharing the SLRA's success stories and best practices with other government departments can serve as a model for effective governance and operational improvements.

Conclusion: The transformation of the SLRA through strategic reforms and dedicated efforts serves as an exemplary model for other government departments. By focusing on autonomy, sustainable funding, comprehensive assessments, and stakeholder engagement, similar success can be achieved in various sectors, thereby promoting good governance and sustainable development throughout Sierra Leone.

Reflection and Engagement Questions

This chapter offers a comprehensive overview of Sierra Leone's governance history, from the colonial era through post-independence to the present day. It highlights the significant milestones and persistent challenges that have shaped the nation's political and social fabric. By understanding this historical context, readers can better appreciate the current governance dynamics and the critical need for effective frameworks that promote stability, inclusivity, and development. The chapter underscores the importance of learning from past experiences to build a stronger, more resilient future for Sierra Leone.

1. **Historical Context:**
   - How has the historical evolution of governance in Sierra Leone influenced the current political and social environment?
   - _____
     _____

2. **Learning from the Past:**
   - What lessons can be drawn from Sierra Leone's post-independence governance challenges and achievements that can inform future reforms?
   - _____
     _____

3. **Role of Advocacy Groups:**
   - In what ways can advocacy groups like SLAM contribute to fostering peace, stability, and effective negotiation in Sierra Leone's political landscape?
   - _____
     _____

4. **Current Challenges:**
   - Reflect on the current governance challenges in Sierra Leone discussed in this chapter. What innovative solutions or strategies can be proposed to address these issues?
   - _____
     _____

5. **Building a Framework:**
    - Considering the historical and contemporary insights provided, what key elements do you believe are essential for establishing a robust governance framework that can sustain Sierra Leone's development? How can you contribute to this process?

    - _____
      _____

CHAPTER 5:
PRINCIPLES OF GOOD GOVERNANCE

Introduction: Sierra Leone, a nation rich in culture and history, stands at a critical juncture in its developmental trajectory. Following a protracted civil war and various governance challenges, the country is steadily making strides toward stability and improved governance. However, significant hurdles remain in fully realizing of democratization and good governance. This context presents a crucial opportunity for SLAM to play a pivotal role as both s research-think tank and implementation partner with other organisation that work on similar initiatives.

SLAM, an organization is dedicated to empowering and improving Sierra Leone through diverse diaspora engagement, is committed to advocating for good governance and democratization. Our mission is to support Sierra Leone's development and ensure that this development is conducted on a foundation of accountability, transparency, and inclusiveness.

This report outlines essential principles of good governance and democracy tailored to the Sierra Leonean context. By advocating for these principles, SLAM aims to contribute to the building of a democratic society in which every citizen can participate meaningfully. These principles serve as benchmarks for our initiatives and guide our efforts to influence policy, educate the citizenry, and engage with governmental and non-governmental stakeholders.

By aligning our work with these principles, SLAM is not just participating in the discourse of development but actively working to mold the governance structures in Sierra Leone to be more reflective of its people's needs and rights. The ensuing sections detail each principle, providing a roadmap for our advocacy and a framework for our actions as we strive to contribute effectively to Sierra Leone's democratic governance and overall development.

Informed by the 1991 Constitution of Sierra Leone, this document outlines the principles of good governance and democratic values that are essential for the nation's development. Key constitutional provisions, such as the Fundamental Principles of State Policy (Chapter II), the Recognition and Protection of Fundamental Human Rights and Freedoms (Chapter III), and the Duties of the Citizen (Chapter II, Section 13), provide a foundational framework for ensuring transparency, accountability, participation, and inclusiveness in governance. This alignment with constitutional mandates ensures that our advocacy and actions are grounded in the legal and ethical standards set forth by the nation's highest legal document.

1. **Accountability**: Fundamental to good governance, accountability ensures that public officials and institutions are answerable for their actions. By promoting accountability, SLAM can help build trust in government and ensure that officials uphold their responsibilities to the public. Constitution Reference: Fundamental Principles of State Policy, Chapter II, Sections 5 and 10.

2. **Transparency**: Transparency makes information freely available and accessible, enabling public scrutiny and participation in governance.

SLAM's role in enhancing transparency can lead to a more informed and engaged citizenry, reduce corruption, and enhance policy effectiveness. Constitution Reference: Fundamental Principles of State Policy, Chapter II, Sections 5 and 10.

3. Responsiveness: This principle requires institutions to react to the public's needs in a timely and appropriate manner. SLAM can advocate for responsiveness in public services and policy adjustments that reflect Sierra Leoneans' current needs, improving public satisfaction and trust in governance. Constitution Reference: Duties of the Citizen, Chapter II, Section 13.

4. Consensus-Oriented: Effective governance mediates differing interests to reach a broad consensus on what is best for the entire community. SLAM can facilitate dialogue and negotiation processes that promote consensus-building, ensuring policies are inclusive and equitably benefit all segments of society. **Constitution Reference**: Fundamental Principles of State Policy, Chapter II, Section 10.

5. Equity and Inclusiveness: Ensuring that all community members feel included, and their interests are considered in the governance process is crucial for fairness and social stability. SLAM's efforts to promote equity and inclusiveness can help ensure no group is left behind, fostering a more cohesive society.

6. Effectiveness and Efficiency: Governance should produce results that meet society's needs while making the best use of available resources. SLAM can help enhance governance's effectiveness and efficiency by promoting better management practices and accountability in resource utilization. **Constitution Reference**: Fundamental Principles of State Policy, Chapter II,

Section 8 and Recognition and Protection of Fundamental Human Rights, Chapter III, Section 27.

7. Rule of Law: This requires that laws be enforced impartially and that all individuals, including those in power, be held accountable under the law. By advocating for the rule of law, SLAM supports a stable legal and political environment where rights are protected, and justice is served equitably. **Constitution Reference:** Recognition and Protection of Fundamental Human Rights, Chapter III, Sections 15 and 16.

8. Participation: All individuals should be able to directly influence decisions that affect their lives or through legitimate intermediaries. SLAM's initiatives to enhance participation can empower citizens and ensure their voices are heard and considered in decision-making processes. **Constitution Reference:** Fundamental Principles of State Policy, Chapter II, Section 2(c).

9. Guide Rails for Upholding Democratic Principles in Sierra Leone: Ensuring the sustainability of democratic governance in Sierra Leone requires a steadfast commitment to core principles, particularly during challenging times. This document outlines essential guidelines to maintain democracy's integrity, emphasizing the importance of long-term goals, inclusivity, accountability, and ethical conduct.

These principles form a robust framework for good governance and democratic functioning. SLAM's advocacy and initiatives aimed at promoting these principles can significantly contribute to Sierra Leone's political, social, and economic development. This framework guides the organization's strategies and aligns with the broader goal of establishing a transparent, accountable, and participatory governance system in Sierra Leone.

Accountability in Governance

Principle: Accountability is a cornerstone of good governance. For democratic institutions to function effectively and maintain public trust, public officials and institutions must be answerable to the citizens they serve. This means that they must take responsibility for their actions and decisions, be transparent in their operations, and be willing to be held accountable for their outcomes.

Context and Importance in Sierra Leone: In the Sierra Leonean context, where historical challenges have sometimes undermined public trust in government, accountability is especially critical. Ensuring that public officials and institutions are accountable can help rebuild trust, enhance government effectiveness, and encourage greater public participation in the democratic process. It can also serve as a deterrent to corruption and misuse of power, issues that have plagued many developing nations.

SLAM's Role in Promoting Accountability: SLAM has a vital role to play in advocating for enhanced accountability. By monitoring government actions, providing platforms for public discourse, and educating citizens on their rights to hold their leaders accountable, SLAM can help ensure that accountability is not just a principle but a practical reality in Sierra Leone's governance.

Strategies for Enhancing Accountability:

- **Public Monitoring Initiatives**: SLAM can organize and support initiatives that monitor government projects and expenditures. This can be achieved through partnerships with local NGOs and community groups trained to observe and report on government activities.

- **Advocacy for Legal Frameworks**: Work towards strengthening the legal frameworks that demand accountability from public officials. This includes advocating for laws that enforce clear and strict reporting requirements for government officials and provide penalties for non-compliance.

- **Educational Programs**: Implement educational programs that inform citizens of their rights and the mechanisms available to hold public officials accountable. Knowledge is power, and an informed citizenry is critical to a functioning democracy.

- **Transparency Platforms**: Develop and promote platforms that provide government data and information access. These platforms can include websites where government documents are published or apps that track legislative activities and provide updates on government initiatives.

- **Engagement in Policy Discussions**: SLAM actively participates in policy discussions and public forums where governance issues are debated. SLAM can influence policy decisions and the public discourse surrounding governance by being a vocal advocate for accountability.

Reference: "Making Accountability Work: Dilemmas for Evaluation and for Audit" by Pollitt, C., et al. (2003)[7] provides a detailed analysis of the complexities of implementing accountability in public administration.

The paper explores how accountability is not merely a technical

---

[7] Making Accountability Work: Dilemmas for Evaluation and for Audit"

requirement but a fundamental aspect that influences the effectiveness and legitimacy of governance. By understanding these challenges and strategies, SLAM can better tailor its advocacy efforts to be more impactful in the Sierra Leon context.

By focusing on these strategies and leveraging the insights from established research, SLAM can significantly contribute to strengthening the accountability of public officials and institutions in Sierra Leone. This, in turn, will help in building a more transparent, accountable, and democratic society.

Transparency in Governance

Principle: Transparency is a vital principle of good governance. It ensures that information about governance policies, practices, and outcomes is available and accessible to all citizens. This openness lets the public know what their government is doing, which fosters greater accountability, trust, and participation in the democratic process.

Context and Importance in Sierra Leone: For Sierra Leone, a country striving to solidify its democratic institutions and heal from past conflicts, transparency is crucial. The public's ability to access clear and accurate information about government actions can help rebuild trust in political processes and institutions. It also empowers citizens to engage actively in governance, from local community decisions to national policy-making.

SLAM's Role in Promoting Transparency: SLAM can play a key role in advocating for and ensuring transparency in several ways. By actively engaging in and promoting transparent practices, SLAM can help set open standards that government bodies are encouraged to follow.

Strategies for Enhancing Transparency:

- **Advocacy for Information Laws**: SLAM should advocate for robust freedom of information laws that compel government agencies to publish data and information about their operations and decisions. This includes pushing for enacting and enforcing laws ensuring public access to governmental meetings, records, and electronic data.

- **Development of Information Platforms**: Develop and support platforms that make governmental data accessible to the public. This could include online portals that host government financial records, decision-making processes, and evaluations of government projects.

- **Training Journalists and Civic Groups**: Train journalists and civic activists in investigative skills and legal rights to information. Equipped with the right tools; these groups can effectively uncover, analyze, and disseminate information about government activities, further promoting a culture of transparency.

- **Public Awareness Campaigns**: Conduct public awareness campaigns to educate citizens on their right to access information and how to exercise this right. Awareness is a critical step in ensuring that transparency measures are utilized by the public.

- **Partnerships with International Transparency Organizations**: Collaborate with global transparency organizations such as Transparency International to adopt international best practices and standards.

This can help SLAM develop globally informed and tested local strategies.

Reference: "Transparency in Politics and the Media: Accountability and Open Government" edited by Nigel Bowles, James T. Hamilton, and David A. L. Levy (2014) discusses the critical role of transparency in building accountable governance.[8] The book explores how transparency not only helps reduce corruption but also enhances public trust and promotes an informed citizenry, which is essential for the functioning of a democratic society.

By implementing these strategies, SLAM can effectively champion the cause of transparency in Sierra Leone, ensuring that the government's actions are visible and understandable to all. This visibility is essential for fostering an environment where citizens feel involved and invested in their governance, contributing to the stability and development of Sierra Leone's democratic institutions.

Responsiveness in Governance

Principle: Responsiveness is a key component of good governance, emphasizing the need for institutions to not only listen to but also act upon the public's needs and concerns of the public. This principle requires that decisions and actions by government and public institutions are aligned with current public needs and implemented promptly and appropriately.

---

[8] Transparency in Politics and the Media: Accountability and Open Government edited by Nigel Bowles, James T. Hamilton, and David A. L. Levy (2014)

Context and Importance in Sierra Leone: In Sierra Leone, where societal needs can change rapidly due to economic fluctuations, environmental challenges, and social issues, the responsiveness of governmental and institutional actions is particularly crucial. Effective responsiveness ensures that policies remain relevant and that public services adapt to meet evolving circumstances, enhancing public satisfaction and trust in government.

SLAM's Role in Promoting Responsiveness: As an advocacy group, SLAM can play a pivotal role in enhancing the responsiveness of public institutions by acting as a bridge between the government and the citizens. SLAM can help articulate the public's needs to the government while also ensuring that the government's intentions and actions are clearly communicated back to the public.

Strategies for Enhancing Responsiveness:

- **Community Engagement Forums**: Organize regular community engagement forums where citizens can directly present their concerns and needs to representatives from relevant governmental and public institutions. These forums provide a direct line of communication from the community to policymakers.

- **Feedback Mechanisms**: Develop and implement effective feedback mechanisms that allow citizens to report on their satisfaction with public services and to log complaints or suggestions for improvements. Mechanisms might include digital platforms, phone hotlines, and regular community surveys.

- **Policy Advocacy**: Advocate for adopting policies that require government agencies to regularly review and adapt their services in response to public feedback and changing needs. This could include legislative frameworks that mandate service quality standards and regular reporting on responsiveness.

- **Capacity Building for Public Officials**: Conduct workshops and training programs for public officials on best practices in responsive governance. Training should focus on empathy, listening skills, and the swift implementation of solutions to public issues.

- **Monitoring and Evaluation Systems**: Help establish robust monitoring and evaluation systems that assess the effectiveness of public services and the extent to which they meet public needs. These systems should provide real-time data to policymakers, helping them make informed decisions that reflect current needs.

Reference: "Public Management and Governance" edited by Tony Bovaird and Elke Loeffler (2016) offers an insightful analysis of how responsiveness is integrated into public service delivery and governance structures.[9] It emphasizes the importance of alignment between public services and the community's needs, suggesting that responsiveness improves service efficiency and strengthens democratic legitimacy and public trust.

---

[9] "Public Management and Governance" edited by Tony Bovaird and Elke Loeffler (2016)

By implementing these strategies, SLAM can significantly enhance the responsiveness of governance in Sierra Leone. This approach will ensure that the governance framework not only listens to but also acts on the public's needs efficiently and effectively, fostering a governance culture that is dynamic, participatory, and tuned to the realities of its citizens.

Consensus-Oriented Governance

Principle: Consensus-oriented governance emphasizes the importance of mediating differing interests within a community or a society to achieve a broad consensus on public issues, policies, and procedures. This principle is vital in ensuring that decisions are made considering diverse viewpoints and lead to equitable and sustainable outcomes that reflect the community's collective best interest.

Context and Importance in Sierra Leone: In Sierra Leone, a country with a diverse population and a history of ethnic and political divisions, fostering consensus-oriented governance is crucial for national cohesion and sustainable development. Consensus-building helps mitigate conflicts, enhances social harmony, and promotes more inclusive and effective policymaking.

SLAM's Role in Promoting Consensus-Oriented Governance: SLAM can significantly promote consensus-oriented governance by facilitating dialogue and understanding among various stakeholders. SLAM's initiatives can help bridge gaps between different groups and foster a culture of collaboration and mutual respect.

Strategies for Enhancing Consensus-Oriented Governance:

- **Facilitate Inclusive Dialogues**: Organize forums and workshops that bring together representatives from diverse groups, including political parties, ethnic groups, civil society organizations, and the private sector. These dialogues should discuss issues, negotiate differences, and formulate agreed-upon solutions.

- **Strengthen Mediation Capabilities**: Train local leaders and officials in consensus-building techniques to develop conflict resolution and mediation capacities. These skills are essential for navigating complex discussions and leading to agreements that reflect the collective interests of all parties.

- **Promote Collaborative Policymaking**: Advocate for and support establishing collaborative policymaking processes where stakeholders from various sectors develop, review, and implement policies. This approach ensures that policies are comprehensive, widely accepted, and supported.

- **Support Civic Education**: Implement civic education programs emphasizing the importance of consensus in democracy. Educate citizens on how their engagement and cooperation can lead to better governance and societal outcomes.

- **Utilize Technology for Broader Engagement**: Deploy digital platforms that facilitate wider community participation in governance processes. These platforms can gather input, conduct polls, and allow for remote participation in forums and discussions, ensuring broader community involvement in consensus-building.

Reference: "Governance: A Very Short Introduction" by Mark Bevir (2012) provides an excellent foundation for understanding how consensus is integral to democratic governance.[10] Bevir argues that consensus does not necessarily mean unanimous agreement but involves processes that seek to reconcile differing needs and preferences to arrive at decisions that are acceptable to the majority.

By adopting these strategies, SLAM can significantly enhance consensus-oriented governance in Sierra Leone. Such efforts will improve the effectiveness of governance and strengthen the nation's social fabric by promoting unity and collective action. This approach will help Sierra Leone achieve a more stable, inclusive, and prosperous future, where policies and decisions are made through broad-based participation and agreement.

Equity and Inclusiveness in Governance

Principle: Equity and inclusiveness ensure that all community members, especially the most vulnerable and marginalized, have their interests represented in the governance process. This principle is essential for creating fair and just policies and helps build a society where everyone has the opportunity to improve their life conditions without discrimination.

Context and Importance in Sierra Leone: Like in many other diverse societies, historical inequalities and ongoing disparities pose significant challenges in Sierra Leone.

---

[10] Governance: A Very Short Introduction by Mark Bevir (2012)

Ensuring that governance processes are equitable and inclusive is crucial for addressing these issues effectively. It's important that all groups, including women, youth, ethnic minorities, and people with disabilities, have their voices heard and their needs addressed in the public domain.

SLAM's Role in Promoting Equity and Inclusiveness: SLAM can advocate for and implement strategies that ensure greater equity and inclusiveness in governance. By doing so, SLAM can help transform Sierra Leone's political landscape into one that champions the rights and interests of all segments of society.

Strategies for Enhancing Equity and Inclusiveness:

- **Advocacy for Inclusive Policies**: SLAM should actively lobby for developing and implementing policies that specifically address the needs of underrepresented and marginalized groups. This involves working closely with policymakers to ensure these groups are considered in all legislative and governance processes.

- **Representation in Decision-Making**: Promote and support the participation of marginalized groups in political and governance structures. This could involve supporting campaigns to increase their political representation in elected offices and public service.

- **Community Outreach Programs**: Implement outreach programs to engage with various community groups to understand their concerns and needs. These programs should serve as a platform for these groups to voice their opinions and influence policy directly.

- **Capacity-Building Initiatives:** Organize workshops and training sessions that empower underrepresented groups with the skills needed to participate effectively in governance processes. Focus on legal rights, advocacy techniques, and public speaking.

- **Monitoring and Reporting**: Regularly monitor and report on the inclusiveness of governance processes. This could involve setting up a watchdog body that evaluates public policies and their impact on different demographic groups to maintain equity.

Social Equity: Social equity focuses on ensuring that all citizens have equal access to resources and opportunities, regardless of their gender, ethnicity, or socio-economic status. This principle is vital for reducing inequality and promoting social justice. Key strategies for achieving social equity include:

- **Equitable Resource** Allocation: Ensuring government resources are distributed fairly across different regions and communities.

- **Access to Services**: Guaranteeing that all citizens have access to essential services such as healthcare, education, and housing.

- **Anti-Discrimination Laws:** Enforcing laws and regulations prohibiting discrimination based on gender, ethnicity, or other characteristics.

- **Economic Opportunities**: Creating programs that provide economic opportunities for disadvantaged groups, including job training and microfinance initiatives.

Reference: "Social Inclusion and the Legal System: Public Interest Law in Ireland" by Ivana Bacik (2002) provides a detailed examination of how legal frameworks can be designed to ensure equity and inclusion.[11] Bacik's analysis highlights the importance of legal structures in facilitating or hindering social inclusion, offering valuable insights that can be adapted to the Sierra Leone context to develop more inclusive governance frameworks.

By implementing these strategies, SLAM will advocate for and actively contribute to a governance process that respects and reflects the diversity of Sierra Leone's population. Such efforts are vital for fostering a sense of belonging and community among all population segments, ultimately contributing to a more stable, cohesive, and prosperous society.

Effectiveness and Efficiency in Governance

Principle: The principles of effectiveness and efficiency require that governance processes and institutions not only achieve their intended outcomes but also make the best possible use of available resources. This means optimizing operations to avoid waste and ensuring that resources are directed toward activities that benefit society most.

Context and Importance in Sierra Leone: For Sierra Leone, a nation with limited resources and significant developmental needs, ensuring effective and efficient governance is particularly crucial.

---

[11] "Social Inclusion and the Legal System: Public Interest Law in Ireland" by Ivana Bacik (2002)

Efficient resource use ensures that more can be done with less, a vital strategy in contexts where fiscal constraints are a constant challenge. On the other hand, effective governance ensures that the services and policies implemented truly meet the population's needs, thereby enhancing public trust and support for government institutions.

SLAM's Role in Promoting Effectiveness and Efficiency: SLAM can play a crucial role in promoting governance that is both effective and efficient. Through advocacy, monitoring, and capacity-building efforts, SLAM can help ensure that governance processes in Sierra Leone are designed and executed to maximize impact and resource utilization.

Strategies for Enhancing Effectiveness and Efficiency:

- **Policy and Process Evaluation**: Advocate for and regularly evaluate policies and governance processes to assess their effectiveness and efficiency. Use findings from these evaluations to push for necessary reforms or improvements.

- **Capacity Building for Public Officials**: Implement training programs for government officials that focus on best practices in public administration, particularly in areas such as project management, resource allocation, and performance monitoring. These skills are essential for enhancing the efficiency of government operations.

- **Promotion of Technology in Governance**: Encourage the adoption of technological solutions that can improve the efficiency of governance processes. This could include digital platforms for service delivery, which reduce costs and increase accessibility, or data management systems that improve decision-making quality.

- **Streamlining Procedures**: Work with government agencies to streamline procedures and reduce bureaucratic red tape. Simplifying processes can significantly increase the speed and reduce the costs of service delivery, making governance more efficient.

- **Transparency and Accountability Mechanisms**: Strengthen transparency and accountability mechanisms to ensure that inefficiencies and ineffectiveness are quickly identified and addressed. This includes supporting the implementation of performance-based accountability systems for public officials and agencies.

Reference: "Managing Effectiveness in the Public Sector" by Ewan Ferlie et al. (2005) provides comprehensive insights into how public sector organizations can enhance their effectiveness and efficiency.[12] The book discusses various strategies, such as performance management, strategic planning, and quality assurance, which are critical for improving the outcomes of public sector operations. These strategies can be tailored to fit the specific context of Sierra Leone, helping to enhance the overall performance of its governance institutions.

By focusing on these strategies, SLAM can contribute significantly to making governance in Sierra Leone not only more responsive and transparent but also more effective and efficient. This dual focus on outcomes and resource utilization is essential for meeting the country's developmental goals in a sustainable manner.

---

[12] "Managing Effectiveness in the Public Sector," (2005) Ewan Ferlie et al.

Rule of Law in Governance

Principle: The rule of law is a cornerstone of good governance and a fundamental principle of a functioning democracy. It requires that all laws are fairly applied and enforced without discrimination. Under the rule of law, every individual, including those in positions of power, is subject to the law. Moreover, it ensures the protection of all human rights, including those of minorities, ensuring that everyone has access to justice and that their rights are protected by transparent and accountable institutions.

Context and Importance in Sierra Leone: In Sierra Leone, the strengthening of the rule of law is crucial for ensuring lasting peace and stability, particularly following a history of civil conflict and challenges with corruption. A strong rule of law promotes fairness and equity, reduces corruption, and helps build public trust in institutions. For Sierra Leone, reinforcing legal frameworks and ensuring their impartial enforcement is not only about maintaining order but also about fostering an environment where economic and social development can flourish.

SLAM's Role in Promoting the Rule of Law: SLAM can advocate for a robust rule of law as part of its broader mission to enhance governance in the country. By focusing on the rule of law, SLAM can help ensure that justice and equity are integral to the nation's governance structures.

Strategies for Enhancing the Rule of Law:

- **Legal Education and Awareness**: Increase legal literacy among citizens to empower them to assert their rights and engage effectively with the legal system.

This involves conducting workshops, distributing informational materials, and leveraging media platforms to educate the public about their legal rights and responsibilities.

- **Advocacy for Judicial Independence**: Advocate for policies and frameworks that ensure the independence of the judiciary. This is crucial for ensuring that legal judgments are made impartially and without influence from other branches of government or external pressures.

- **Support for Legal Reforms**: Collaborate with legal experts, civil society organizations, and international partners to propose and support reforms that enhance legal frameworks. These reforms should aim to make the legal system more accessible, transparent, and responsive to the needs of all segments of society, including minorities.

- **Monitoring and Reporting on Legal Enforcement**: Establish mechanisms to monitor and report on the enforcement of laws and the functioning of the judiciary. This could involve setting up observer missions in courts or conducting periodic reviews of legal proceedings to ensure that laws are applied fairly and efficiently.

- **Protection of Minority Rights**: Specifically, focus on the legal protection of minorities by advocating for laws that protect against discrimination and inequality. Engage in legal aid initiatives to assist minorities in navigating the legal system to defend their rights.

Reference: "Rule of Law Reform and Development: Charting the Fragile Path of Progress" by Michael J. Trebilcock and Ronald J. Daniels (2008) provides a thorough exploration of the role of the rule of law in supporting democratic governance and socio-economic development.[13] The authors discuss the necessity of legal reform in post-conflict societies and the ways in which to strengthened legal systems can contribute to broader governance reforms. The insights from this book can guide SLAM in advocating for and implementing strategies that fortify the rule of law in Sierra Leone.

By adopting these strategies, SLAM can play a pivotal role in enhancing the rule of law in Sierra Leone. A strong rule of law not only secures justice for all citizens but also lays the foundation for peace, stability, and development in the nation. This comprehensive approach ensures that the rule of law is not only a principle upheld in books but a reality experienced by all Sierra Leoneans in their daily lives.

Participation in Governance

Principle: Participation is a fundamental aspect of democracy and good governance. It ensures that all men and women have the ability to influence decision-making processes, either directly or through representatives who articulate their interests. True participatory governance fosters inclusion, legitimacy, and consensus, allowing policies to reflect the diverse needs and aspirations of the entire community.

---

[13] "Rule of Law Reform and Development: Charting the Fragile Path of Progress" (2008) by Michael J. Trebilcock and Ronald J. Daniels.

Context and Importance in Sierra Leone: For Sierra Leone, enhancing participation is crucial in transitioning from a post-conflict society to one marked by stable and inclusive governance. Engaging all segments of society—including women, youth, and marginalized groups—in the political process helps to ensure that different perspectives are considered and everyone has a stake in the country's future. Increased participation can lead to more resilient and effective policies that are widely supported and more likely to succeed.

SLAM's Role in Promoting Participation: SLAM can play a crucial role in promoting widespread participation in governance processes. By advocating for and facilitating broader inclusion, SLAM can help bridge gaps between the government and its citizens, ensuring that governance truly reflects the will of the people.

Strategies for Enhancing Participation:

- **Civic Education Programs**: Implement comprehensive civic education programs that inform citizens about their rights, responsibilities, and the importance of their involvement in governance. These programs can cover how to vote, the significance of civic duties, and ways to engage with and influence local and national policy.

- **Support for Civil Society Organizations**: Strengthen civil society organizations that act as intermediaries between the government and the citizens. Provide these organizations with the resources, training, and platforms needed to effectively represent their constituents' interests.

- **Facilitation of Public Forums and Consultations**: Organize and facilitate regular public forums, town hall meetings, and consultations on key legislative and policy issues. Ensure these forums are accessible to a broad segment of the population to gather diverse viewpoints and foster a culture of open dialogue.

- **Use of Digital Platforms to Enhance Participation**: Leverage technology to create more participatory opportunities. This could include online surveys, virtual town halls, and social media campaigns that allow people to voice their opinions, vote on issues, and engage in discussions from anywhere in the country.

- **Inclusive Policy-Making Processes**: Advocate for and help implement mechanisms that ensure policies are developed with direct input from all stakeholders. This can involve setting up advisory panels that include representatives from various societal groups, ensuring their voices are heard in the policy-making process.

Reference: "Voice and Participation in Global Governance: From Local Struggles to Global Advocacy" by Thomas Pogge and Akmal Qureshi (2012) discusses the importance of enhancing participatory governance through various mechanisms.[14] The authors explore how increased participation at local and global levels can lead to more equitable and effective governance. Drawing on examples from around the world, the book offers valuable insights into how different communities have successfully increased public involvement in governance.

---

[14] "Voice and Participation in Global Governance: From Local Struggles to Global Advocacy" (2012), Thomas Pogge and Akmal Qureshi.

By following these strategies, SLAM can help ensure that Sierra Leone's governance processes are participatory and inclusive. This will not only improve the quality and effectiveness of governance but also help build a more cohesive and democratic society. Engaging all citizens in decision-making processes strengthens democratic institutions and fosters a sense of ownership and responsibility among the populace, contributing to the overall stability and development of Sierra Leone.

Reflection and Engagement Questions

This Chapter delves into the core principles of good governance that are vital for the development and stability of Sierra Leone. These principles include accountability, transparency, responsiveness, consensus orientation, equity and inclusiveness, effectiveness and efficiency, rule of law, and participation. Each principle is explored in the context of Sierra Leone's unique political and social landscape, providing a comprehensive understanding of their importance and application.

The chapter emphasizes that good governance is not just about implementing policies but about creating an environment where citizens feel empowered, informed, and involved in the decision-making processes. By adhering to these principles, Sierra Leone can build a governance framework that is resilient, just, and conducive to sustainable development.

Readers are encouraged to reflect on how these principles can be practically applied in their communities and what role they can play in advocating for these standards.

Building A Nation

Through active participation, informed advocacy, and a commitment to democratic values, every citizen can contribute to the realization of good governance in Sierra Leone.

1. **Understanding Accountability:**
   - How does accountability in governance impact public trust and government effectiveness? Reflect on ways accountability mechanisms can be strengthened in Sierra Leone.

   - _____
     _____

2. **Promoting Transparency:**
   - Why is transparency essential for good governance? Discuss the role of open data portals and public access to government information in promoting transparency.

   - _____
     _____

3. **Ensuring Responsiveness:**
   - In what ways can the government become more responsive to the needs and concerns of its citizens? Consider the importance of public consultations and feedback mechanisms.

   - _____
     _____

4. **Fostering Inclusivity and Equity:**

    o   How can inclusivity and equity be promoted in governance to ensure that all segments of society are represented? Reflect on the role of policies that support marginalized groups.

    o   _____
    _____

5. **Upholding the Rule of Law:**

    o   What are the key components of the rule of law, and why are they crucial for good governance? Discuss the importance of an independent judiciary and effective law enforcement agencies.

    _____
    _____

CHAPTER 6:
KEY STRATEGIES FOR ENHANCING GOVERNANCE

Public Monitoring Initiatives: Public monitoring initiatives are a vital component in promoting good governance and ensuring accountability in Sierra Leone. These initiatives empower citizens to actively engage in oversight of government activities, thereby enhancing transparency and reducing corruption. By involving the public in monitoring processes, governments can be held accountable for their actions, leading to more responsible and effective governance. This section explores the significance of public monitoring initiatives and provides practical examples and strategies for their implementation.

Significance of Public Monitoring Initiatives

1. **Empowerment of Citizens:**
    - Public monitoring initiatives empower citizens by providing them with the tools and knowledge necessary to oversee government activities. This empowerment fosters a sense of ownership and responsibility among citizens, encouraging active participation in governance.

2. **Increased Transparency:**
   - These initiatives enhance transparency by making government operations more visible to the public. When citizens have access to information about government activities, they can better understand and scrutinize decisions and actions, leading to increased accountability.

3. **Reduction of Corruption:**
   - By involving citizens in monitoring government activities, public monitoring initiatives can help reduce corruption. Citizens can report irregularities and misconduct, deterring public officials from engaging in corrupt practices.

4. **Enhanced Public Trust:**
   - Transparency and accountability fostered by public monitoring initiatives can lead to increased public trust in government institutions. When citizens see that their concerns are addressed and that the government is accountable, they are more likely to trust and support public institutions.

Practical Examples of Public Monitoring Initiatives

1. **Citizen Report Cards:**
   - Citizen report cards are tools used to collect feedback from citizens about the quality of public services.

These report cards can highlight areas where services are lacking and provide data for advocacy efforts to improve service delivery.

2. **Community Scorecards:**
    - Community scorecards are participatory tools that enable communities to assess and provide feedback on public services. These scorecards can be used to facilitate dialogue between service providers and the community, leading to improvements in service delivery.

3. **Social Audits:**
    - Social audits involve the examination of government records, policies, and activities by the public to ensure transparency and accountability. These audits can uncover discrepancies and irregularities, prompting corrective actions.

4. **Public Expenditure Tracking Surveys (PETS):**
    - PETS are used to track the flow of public funds from the government to the intended beneficiaries. These surveys can identify bottlenecks and leakages in the system, ensuring that resources are used effectively.

Strategies for Implementing Public Monitoring Initiatives

1. **Capacity Building:**
   - Building the capacity of citizens and civil society organizations (CSOs) is crucial for the success of public monitoring initiatives. Training programs can equip individuals and organizations with the skills needed to effectively monitor and report on government activities.

2. **Partnerships with CSOs:**
   - Collaborating with CSOs can enhance the reach and impact of public monitoring initiatives. CSOs can provide valuable support in organizing and implementing these initiatives, as well as advocating for changes based on their findings.

3. **Use of Technology:**
   - Leveraging technology can facilitate the collection and dissemination of information for public monitoring initiatives. Mobile applications, online platforms, and social media can be used to gather feedback, report irregularities, and share information with a broader audience.

4. **Government Engagement:**
   - Engaging with government officials and institutions is essential for the success of public monitoring initiatives. Building constructive relationships can help ensure that the findings from these initiatives are taken seriously and lead to meaningful changes.

5. **Public Awareness Campaigns:**
    - Raising awareness about the importance of public monitoring and how citizens can get involved is critical. Public awareness campaigns can inform citizens about their rights and the mechanisms available for them to participate in monitoring government activities.

Conclusion: Public monitoring initiatives are powerful tools for enhancing governance in Sierra Leone. By empowering citizens, increasing transparency, reducing corruption, and building public trust, these initiatives can significantly contribute to more accountable and effective governance. Implementing public monitoring initiatives requires capacity building, partnerships with CSOs, the use of technology, engagement with government, and public awareness campaigns. Through these efforts, Sierra Leone can foster a culture of transparency and accountability, leading to improved governance and better outcomes for its citizens.

Advocacy for Legal Frameworks: Advocating for robust legal frameworks is essential for establishing and maintaining good governance in Sierra Leone. Effective legal frameworks provide the foundation for transparency, accountability, and the protection of citizens' rights. They create the legal basis for public monitoring, anti-corruption measures, and other governance initiatives. This section delves into the importance of advocating for legal frameworks, highlights successful examples, and outlines strategies for effective advocacy.

Significance of Legal Frameworks

1. **Foundation for Governance:**
   - Legal frameworks serve as the foundation for governance, outlining the rules, regulations, and standards that govern the actions of public officials and institutions. They ensure that governance processes are consistent, transparent, and fair.

2. **Protection of Rights:**
   - Robust legal frameworks protect the rights of citizens, ensuring that they can participate in governance processes, access information, and hold public officials accountable. This protection is crucial for fostering a democratic society.

3. **Enhancement of Accountability:**
   - Legal frameworks establish mechanisms for accountability, such as audits, oversight bodies, and reporting requirements. These mechanisms help ensure that public officials are held responsible for their actions and decisions.

4. **Reduction of Corruption:**
   - Anti-corruption laws and regulations are integral components of legal frameworks. By defining and penalizing corrupt practices, these laws deter misconduct and promote ethical behavior in public service.

Successful Examples of Legal Framework Advocacy

1. **Freedom of Information Laws:**
   - Advocacy for freedom of information laws has been successful in many countries, leading to greater transparency and public access to government information. These laws empower citizens to request and receive information about government activities, enhancing accountability.

2. **Anti-Corruption Legislation:**
   - Anti-corruption laws and regulations have been advocated for and implemented in various countries, resulting in the establishment of anti-corruption commissions and the prosecution of corrupt officials. These laws help create a culture of integrity and reduce corruption.

3. **Judicial Reforms:**
   - Advocacy for judicial reforms has led to the strengthening of judicial independence and the establishment of fair and impartial legal systems. Strong judicial frameworks ensure that laws are enforced consistently and that citizens have access to justice.

4. **Electoral Reforms:**
   - Electoral reforms, such as the implementation of transparent voting processes and independent electoral commissions, have been successfully advocated for in several countries.

These reforms enhance the credibility and fairness of elections, promoting democratic governance.

Strategies for Effective Advocacy

1. **Building Coalitions:**
   - Forming coalitions with civil society organizations (CSOs), community groups, and other stakeholders can amplify advocacy efforts. Coalitions can leverage their collective resources and influence to push for legal reforms.

2. **Engaging Policymakers:**
   - Building relationships with policymakers and legislators is crucial for effective advocacy. Engaging in dialogue, providing evidence-based recommendations, and highlighting the benefits of proposed legal reforms can help gain their support.

3. **Public Awareness Campaigns:**
   - Raising public awareness about the importance of legal frameworks and the specific reforms being advocated for is essential. Public support can pressure policymakers to take action and create a groundswell of demand for change.

4. **Research and Evidence:**
   - Conducting research and gathering evidence to support advocacy efforts is vital.

Well-researched reports, case studies, and data can strengthen the case for legal reforms and provide a solid foundation for advocacy campaigns.

5. **Leveraging Media:**
    - Utilizing media platforms to highlight governance issues and advocate for legal reforms can increase visibility and public engagement. Op-eds, news articles, social media campaigns, and public service announcements can disseminate information and rally support.

6. **International Partnerships:**
    - Partnering with international organizations and leveraging global networks can provide additional support and resources for advocacy efforts. International partnerships can also bring attention to local issues and help hold governments accountable to international standards.

Conclusion: Advocating for robust legal frameworks is a critical strategy for enhancing governance in Sierra Leone. By establishing the legal basis for transparency, accountability, and the protection of citizens' rights, these frameworks create the conditions necessary for good governance. Successful examples of legal framework advocacy, such as freedom of information laws and anti-corruption legislation, demonstrate the impact of these efforts. Effective advocacy requires building coalitions, engaging policymakers, raising public awareness, conducting research, leveraging media, and forming international partnerships.

Through these strategies, Sierra Leone can develop and implement legal frameworks that promote transparency, accountability, and democratic governance, leading to a more just and equitable society.

Educational Programs: Educational programs play a crucial role in fostering good governance by informing and empowering citizens about their rights, responsibilities, and the mechanisms available for participating in governance processes. These programs can build a more informed and engaged populace capable of holding public officials accountable and contributing to the development of a democratic society. This section explores the importance of educational programs, highlights successful examples, and provides strategies for implementing effective educational initiatives.

Significance of Educational Programs

1. **Informed Citizenry:**
    - Educational programs provide citizens with the knowledge they need to understand governance processes, their rights, and how they can influence public policy. An informed citizenry is better equipped to engage in civic activities and advocate for positive change.

2. **Empowerment:**
    - By educating citizens about their rights and responsibilities, educational programs empower them to take an active role in governance.

This empowerment leads to greater participation in decision-making processes and strengthens democratic institutions.

3. **Promotion of Accountability:**
   o Educated citizens are more likely to hold public officials accountable for their actions. Awareness of accountability mechanisms, such as audits and oversight bodies, enables citizens to demand transparency and integrity from their government.

4. **Reduction of Corruption:**
   o Education about the negative impacts of corruption and the importance of ethical behavior in public service can help reduce corrupt practices. Citizens who understand their rights are more likely to report corruption and advocate for anti-corruption measures.

Successful Examples of Educational Programs

1. **Civic Education Initiatives:**
   o Civic education programs that teach citizens about the structure and function of government, their rights and responsibilities, and how to engage in the political process have been successful in many countries. These programs often include workshops, seminars, and informational materials.

2. **School Curricula:**
    - Integrating governance and civic education into school curricula helps instill democratic values and principles in young people from an early age. These programs can include lessons on the importance of participation, accountability, and transparency.

3. **Public Awareness Campaigns:**
    - Public awareness campaigns that use media, social media, and community outreach to educate citizens about governance issues have proven effective. These campaigns can raise awareness about specific topics, such as anti-corruption efforts or the importance of voting.

4. **Workshops and Seminars:**
    - Workshops and seminars for adults can provide in-depth education on governance topics. These events can be tailored to specific audiences, such as community leaders, journalists, or civil society organizations, to build their capacity to engage in governance processes.

Strategies for Implementing Educational Programs

1. **Collaboration with Educational Institutions:**
    - Partnering with schools, universities, and other educational institutions can help integrate governance education into existing curricula.

These partnerships can also provide resources and expertise for developing educational materials.

2. **Engaging Civil Society Organizations:**
   - Civil society organizations (CSOs) can play a key role in delivering educational programs, as they often have the reach and trust of local communities, making them effective partners in educating citizens about governance.

3. **Utilizing Media and Technology:**
   - Leveraging media and technology can expand the reach of educational programs. Online courses, webinars, social media campaigns, and mobile applications can provide accessible and flexible learning opportunities for a wide audience.

4. **Community-Based Approaches:**
   - Implementing community-based educational programs can ensure that the content is relevant and tailored to the specific needs of the community. Community leaders and local organizations can help facilitate these programs and encourage participation.

5. **Monitoring and Evaluation:**
   - Monitoring and evaluating educational programs is essential for assessing their effectiveness and making improvements.

Feedback from participants, assessment of learning outcomes, and impact evaluations can help refine and enhance these programs.

6. **Government Support:**
    - Securing support from government agencies can provide legitimacy and resources for educational programs. Governments can endorse and promote these initiatives, integrate them into public education systems, and provide adequate funding.

Conclusion: Educational programs are a vital strategy for enhancing governance in Sierra Leone. By informing and empowering citizens, these programs foster a more engaged and accountable populace capable of contributing to democratic processes and holding public officials accountable. Successful examples of educational programs, such as civic education initiatives and public awareness campaigns, demonstrate their potential impact. Effective implementation requires collaboration with educational institutions, engagement with CSOs, utilization of media and technology, community-based approaches, monitoring and evaluation, and government support. Through these strategies, educational programs can promote transparency, accountability, and active participation in governance, leading to a more informed and democratic society.

Transparency Platforms: Transparency platforms are essential tools for promoting good governance by making government operations visible and accessible to the public. These platforms enhance accountability, deter corruption, and foster trust between citizens and their government.

By providing open access to information, transparency platforms enable citizens to monitor government activities, make informed decisions, and engage more effectively in governance processes. This section explores the importance of transparency platforms, highlights successful examples, and outlines strategies for their implementation.

Significance of Transparency Platforms

1. **Enhanced Accountability:**
   - Transparency platforms allow citizens to scrutinize government actions and decisions, holding public officials accountable. This increased oversight helps ensure that officials act in the public interest and adhere to ethical standards.

2. **Reduction of Corruption:**
   - By making government activities transparent, these platforms reduce opportunities for corruption. When information about government operations is readily available, it is harder for corrupt practices to go unnoticed.

3. **Public Trust:**
   - Transparency fosters trust between citizens and their government. When the public can see and understand government operations, they are more likely to trust that officials are working in their best interests.

4. **Informed Decision-Making:**
    - Access to information empowers citizens to make informed decisions about their involvement in governance processes. Whether voting, participating in public consultations, or advocating for policy changes, well-informed citizens are more effective participants.

Successful Examples of Transparency Platforms

1. **Open Data Portals:**
    - Open data portals provide public access to a wide range of government data, including budgets, expenditures, and procurement records. These portals enhance transparency by making it easy for citizens to find and use government information.

2. **Public Procurement Platforms:**
    - Platforms that disclose information about public procurement processes help ensure transparency and fairness in the awarding of government contracts. By making procurement data publicly available, these platforms deter corruption and promote competitive bidding.

3. **Government Websites and Dashboards:**
    - Government websites and online dashboards that provide real-time information about government activities, such as project progress and service delivery metrics, enhance transparency and accountability.

4. **Mobile Applications:**
    - Mobile applications can facilitate transparency by providing citizens with easy access to government information and services. These apps can include features for reporting issues, accessing public records, and receiving updates on government projects.

Strategies for Implementing Transparency Platforms

1. **Developing User-Friendly Interfaces:**
    - Transparency platforms should be designed with user-friendly interfaces that make it easy for citizens to access and understand information. Clear navigation, search functions, and visual aids like charts and graphs can enhance usability.

2. **Ensuring Data Accuracy and Timeliness:**
    - For transparency platforms to be effective, the data provided must be accurate, up-to-date, and comprehensive. Regular updates and stringent data quality controls are essential to maintaining the reliability of these platforms.

3. **Promoting Awareness and Accessibility:**
    - Public awareness campaigns can inform citizens about the availability and benefits of transparency platforms. Ensuring accessibility, including considerations for language, literacy levels, and internet access, is also critical.

4. **Collaborating with Civil Society and Media:**
    - Partnering with civil society organizations (CSOs) and the media can help amplify the reach and impact of transparency platforms. CSOs and media outlets can use the data to conduct analyses, raise public awareness, and advocate for reforms.

5. **Building Government Support:**
    - Securing buy-in from government agencies is crucial for the successful implementation of transparency platforms. Government support can facilitate access to data, provide necessary resources, and promote the use of these platforms.

6. **Leveraging Technology:**
    - Utilizing advanced technologies, such as blockchain for secure and transparent record-keeping or artificial intelligence for data analysis, can enhance the functionality and effectiveness of transparency platforms.

Conclusion: Transparency platforms are powerful tools for enhancing governance in Sierra Leone. By providing open access to information, these platforms promote accountability, reduce corruption, and foster public trust. Successful examples, such as open data portals and public procurement platforms, demonstrate their potential impact. Effective implementation requires developing user-friendly interfaces, ensuring data accuracy and timeliness, promoting awareness and accessibility, collaborating with CSOs and media, building government support, and leveraging technology. Through these strategies, transparency platforms can significantly contribute to a more accountable, transparent, and democratic governance system in Sierra Leone.

Engagement in Policy Discussions: Engaging citizens in policy discussions is a fundamental aspect of democratic governance. This involvement ensures that government policies reflect the needs and preferences of the populace, fostering a sense of ownership and participation among citizens. Active engagement in policy discussions can lead to more effective and equitable policies, enhance public trust, and strengthen democratic institutions. This section explores the importance of citizen engagement in policy discussions, highlights successful examples, and provides strategies for effective implementation.

Significance of Engagement in Policy Discussions

1. **Inclusive Decision-Making:**
   - Engaging citizens in policy discussions ensures that diverse perspectives and needs are considered.

   This inclusivity leads to more comprehensive and equitable policies that address the concerns of all segments of society.

2. **Enhanced Legitimacy:**
   - Policies developed through participatory processes are more likely to be accepted and supported by the public. Engagement fosters a sense of ownership and legitimacy, making it easier to implement and sustain policies.

3. **Improved Policy Outcomes:**
   - Citizen input can provide valuable insights and information that policymakers may not have. This input can lead to better-informed decisions and more effective policy outcomes.

4. **Public Trust and Accountability:**
   - Engaging citizens in policy discussions builds trust between the government and the public. It demonstrates a commitment to transparency and accountability, reinforcing the democratic principle that governance should serve the people.

Successful Examples of Engagement in Policy Discussions

1. **Public Consultations:**
   - Public consultations are forums where citizens can voice their opinions on proposed policies. These consultations can take the form of town hall meetings, public hearings, or online forums, allowing for broad participation.

2. **Participatory Budgeting:**
   - Participatory budgeting involves citizens in the decision-making process regarding the allocation of public funds. This approach allows communities to prioritize spending based on their needs and preferences.

3. **Citizen Assemblies:**
   - Citizen assemblies bring together a representative group of citizens to discuss and deliberate on specific policy issues. These assemblies can provide detailed recommendations to policymakers based on collective input.

4. **Policy Advocacy Groups:**
   - Policy advocacy groups, including civil society organizations (CSOs) and grassroots movements, play a crucial role in representing the interests of various communities in policy discussions. They can mobilize citizens, provide expertise, and advocate for specific policy changes.

Strategies for Effective Engagement in Policy Discussions

1. **Creating Accessible Platforms:**
   - Developing accessible platforms for engagement, such as online portals, mobile apps, and community centers, ensures that all citizens can participate in policy discussions. These platforms should be user-friendly and inclusive.

2. **Conducting Outreach and Education:**
   - Public outreach and education campaigns can inform citizens about ongoing policy discussions and how they can get involved. Educating the public about the importance of their participation can increase engagement.

3. **Facilitating Inclusive Participation:**
   - Ensuring that engagement processes are inclusive and representative of all societal groups, including marginalized and underrepresented communities, is essential. This can be achieved through targeted outreach and by removing barriers to participation.

4. **Providing Feedback Mechanisms:**
   - Implementing mechanisms for providing feedback to participants about how their input was used in policy-making can enhance trust and encourage continued engagement. Transparency about the impact of citizen contributions is crucial.

5. **Building Partnerships:**
    - Collaborating with CSOs, community leaders, and other stakeholders can strengthen engagement efforts. These partners can help mobilize citizens, facilitate discussions, and provide resources and support.

6. **Leveraging Technology:**
    - Utilizing digital tools and technologies can expand the reach and efficiency of engagement efforts. Online surveys, virtual town halls, and social media platforms can facilitate wider participation and real-time feedback.

Conclusion: Engaging citizens in policy discussions is vital for enhancing governance in Sierra Leone. Inclusive decision-making processes lead to more effective and equitable policies, build public trust, and strengthen democratic institutions. Successful examples, such as public consultations and participatory budgeting, demonstrate the positive impact of citizen engagement. Effective strategies for implementing engagement in policy discussions include creating accessible platforms, conducting outreach and education, facilitating inclusive participation, providing feedback mechanisms, building partnerships, and leveraging technology. Through these efforts, Sierra Leone can foster a more participatory and responsive governance system, ensuring that policies reflect the needs and aspirations of its citizens.

Reflection and Engagement Questions

In this chapter, we explore various strategies crucial for enhancing governance in Sierra Leone. These strategies include public monitoring initiatives, advocacy for legal frameworks, educational programs, transparency platforms, and engagement in policy discussions. By implementing these strategies, Sierra Leone can foster a more accountable, transparent, and responsive government. This chapter aims to equip readers with practical tools and approaches to improve governance and promote democratic principles. It underscores the importance of collaboration between government, civil society, and the diaspora to achieve meaningful and sustainable change.

1. **Public Monitoring Initiatives:**

    o How can public monitoring initiatives empower citizens to hold the government accountable? Provide examples of successful public monitoring efforts.

    o _____
    _____

2. **Advocacy for Legal Frameworks:**

    o What legal reforms are necessary to strengthen governance in Sierra Leone? How can advocacy groups influence these reforms?

    o _____
    _____

3. **Educational Programs:**
    - Why are educational programs critical for promoting good governance? Reflect on ways to implement effective educational initiatives in your community.

    - _____
      _____

4. **Transparency Platforms:**
    - How can transparency platforms enhance government accountability and citizen engagement? Discuss the potential impact of digital tools in promoting transparency.

    - _____
      _____

5. **Engagement in Policy Discussions:**
    - What strategies can be used to encourage active participation in policy discussions among Sierra Leoneans? Reflect on the importance of inclusive dialogue in shaping governance policies.

    - _____
      _____

These questions are designed to help readers deeply engage with the chapter's content, fostering critical thinking and encouraging practical application of the strategies discussed to enhance governance in Sierra Leone.

CHAPTER 7:
GUIDE RAILS FOR UPHOLDING DEMOCRATIC PRINCIPLES IN SIERRA LEONE

Introduction: Implementing democratic principles requires commitment, vigilance, and an understanding of the inherent challenges. While many governments have started with these principles and meant well, their failure often lies in the lack of guidelines to protect and uphold democracy. Democracy is inherently difficult, but the rewards are significant in the long term. It is crucial for members to focus on the long-term goals rather than the short-term pressures, especially during tense moments. This section provides guidelines to help maintain these principles and prevent the erosion of democracy.

1. Upholding Democratic Principles

Importance of Adherence: Democracy thrives on the active participation of all citizens, equitable representation, and adherence to the rule of law. It is crucial to continually remind ourselves of these principles to prevent the temptation to sideline them during crises, which can lead to a slippery slope where recovery becomes increasingly difficult. The emphasis should be on the long-term benefits of democracy over short-term expediencies.

Guidelines:

- **Regular Reviews**: Establish a schedule for regular reviews of democratic practices within the organization and governance structures. These reviews should assess adherence to democratic principles and identify areas for improvement.

- **Education and Awareness**: Conduct ongoing education programs for all members to reinforce the importance of democratic principles. This includes workshops, seminars, and discussions on the value of democracy and the consequences of deviating from these principles.

- **Transparency Mechanisms**: Implement transparent decision-making processes that are open to scrutiny. Ensure that all decisions are documented and accessible to the public to maintain trust and accountability.

2. Inclusivity and Participation

Long-Term vs. Short-Term Goals: Democratic adherence may be difficult initially, but it captures diverse ideas and fosters innovation and inclusion, leading to a more resilient society. Members should always consider the long-term impacts of their actions and decisions.

Guidelines:

- **Inclusive Policies**: Develop and enforce policies that ensure the inclusion of diverse groups in decision-making processes. This includes women, youth, minorities, and other marginalized groups.

- **Participation Platforms**: Create platforms for active citizen participation, such as public forums, online consultations, and participatory budgeting processes. These platforms should be easily accessible to all citizens.

- **Feedback Loops**: Establish feedback mechanisms where citizens can voice their opinions and concerns, ensuring these are taken into account in governance processes.

3. Accountability and Transparency

Building Trust: Accountability and transparency are cornerstones of good governance. They ensure that public officials are answerable for their actions and that decisions are made in an open manner.

Guidelines:

- **Public Monitoring**: Organize and support public monitoring initiatives that track government actions and expenditures. Encourage the involvement of civil society organizations in these efforts.

- **Reporting Requirements**: Advocate for strict reporting requirements for government officials and institutions. Ensure that there are clear consequences for non-compliance.

- **Transparent Data**: Promote the availability of government data and information to the public through digital platforms and open data initiatives.

4. Conflict Resolution and Consensus Building

Handling Tense Moments: In times of difficulties, it is crucial to come together and find common ground rather than isolating each other. Consensus-building fosters unity and effective governance.

Guidelines:

- **Dialogue Facilitation**: Organize regular dialogues between government officials, civil society, and citizens to discuss and resolve conflicts. These dialogues should aim for consensus and mutual understanding.

- **Mediation Skills**: Train leaders and officials in mediation and conflict resolution skills to handle disputes effectively and constructively.

- **Collaborative Decision-Making**: Implement decision-making processes that require input and agreement from multiple stakeholders, ensuring that diverse perspectives are considered.

5. Integrity and Ethical Conduct

Maintaining Ethical Standards: Ethical conduct and integrity are essential for maintaining public trust and ensuring the effectiveness of governance.

Guidelines:

- **Code of Conduct**: Develop and enforce a code of conduct for all members and public officials.

This code should outline expected behaviors and the consequences of unethical actions.

- **Whistleblower Protections**: Establish secure channels for reporting unethical behavior without fear of retaliation. Ensure that whistleblowers are protected, and their concerns are addressed promptly.

- **Ethics Training**: Provide regular training on ethical decision-making and integrity to all members and public officials. Emphasize the importance of these principles in maintaining a healthy democracy.

Conclusion: Adhering to these guide rails will help uphold democratic principles in Sierra Leone, even during challenging times. By focusing on long-term goals, fostering inclusivity and participation, ensuring accountability and transparency, building consensus, and maintaining ethical standards, Sierra Leone can build a robust and resilient democratic society. Regular reviews, ongoing education, transparency mechanisms, and inclusive policies are essential. Encouraging public participation, establishing feedback loops, and supporting conflict resolution are critical. Together, these efforts will enhance governance, strengthen public trust, and promote sustainable development, leading to a more inclusive and prosperous nation. Commitment to these values is crucial for a democratic future.

Reflection and Engagement Questions

This chapter focuses on the essential guidelines and frameworks necessary to uphold democratic principles in Sierra Leone. It emphasizes the importance of commitment, vigilance, and understanding the inherent challenges in maintaining a democracy. The chapter outlines practical steps to ensure long-term goals, foster inclusivity, ensure accountability and transparency, build consensus, and maintain ethical standards. By adhering to these guide rails, Sierra Leone can strengthen its democratic institutions and create a more resilient and equitable society.

1. **Long-term Goals:**
    - How can focusing on long-term goals rather than short-term gains help strengthen democratic principles in Sierra Leone?
    - _____

2. **Inclusivity in Governance:**
    - What strategies can be implemented to ensure the inclusion of diverse groups in Sierra Leone's governance processes?
    - _____

3. **Accountability and Transparency:**
    - Reflect on the current mechanisms for accountability and transparency in Sierra Leone.

How can these be improved to build greater public trust in government institutions?

- _____
_____

4. **Conflict Resolution and Consensus Building:**

   - In what ways can Sierra Leone improve its approaches to conflict resolution and consensus building to foster unity and effective governance?

   - _____
   _____

5. **Ethical Standards:**

   - Why is maintaining high ethical standards crucial for public officials, and how can these standards be enforced to prevent misconduct and corruption?

   - _____
   _____

These questions are designed to help readers deeply reflect on the chapter's content, encouraging them to consider practical applications and personal contributions to upholding democratic principles in Sierra Leone.

CHAPTER 8:
DETRIMENTAL PRINCIPLES TO DEMOCRATIC GOVERNANCE IN SIERRA LEONE

Introduction: Sierra Leone, a nation rich in culture and history, stands at a critical juncture in its developmental trajectory. Following a protracted civil war and various governance challenges, the country is steadily making strides toward stability and improved governance. However, significant hurdles remain in fully realizing the principles of democracy and good governance. These challenges are exacerbated by several detrimental principles that undermine the democratic process and hinder national development.

SLAM, an organization dedicated to empowering and improving Sierra Leone through diaspora engagement, is committed to advocating for good governance and democratic principles. Our mission is to support Sierra Leone's development and ensure that this development is conducted on a foundation of accountability, transparency, and inclusiveness.

This report outlines the detrimental principles that obstruct democratic governance in Sierra Leone, referencing the 1991 Constitution of Sierra Leone to highlight the incongruence of these principles with the nation's foundational legal and ethical standards. By identifying and addressing these harmful principles, SLAM aims to contribute to building a democratic society in which every citizen can participate meaningfully. These detrimental principles serve as benchmarks for our advocacy and guide our efforts to influence policy, educate the citizenry, and engage with governmental and non-governmental stakeholders.

Informed by the 1991 Constitution of Sierra Leone, this document outlines the detrimental principles that could be eradicated to foster a democratic society. Key constitutional provisions, such as the Fundamental Principles of State Policy (Chapter II), the Recognition and Protection of Fundamental Human Rights and Freedoms (Chapter III), and the Duties of the Citizen (Chapter II, Section 13), provide a foundational framework for ensuring transparency, accountability, participation, and inclusiveness in governance. This alignment with constitutional mandates ensures that our advocacy and actions are grounded in the legal and ethical standards set forth by the nation's highest legal document.

List of Detrimental Principles

1. **Corruption and Self-Interest ("Where they tie cow that is where they eat")**: The misuse of public resources for personal gain.

2. **Nepotism and Favoritism**: Prioritizing family, friends, or close associates over qualified individuals.

3. **Regionalism and Ethnocentrism**: Focusing on the well-being of one's own region or ethnic group while neglecting others.

4. **Political Partisanship**: Placing party loyalty above national interest.

5. **Patronage Politics**: Distributing public resources and favors in exchange for political support.

6. **Disregard for the Rule of Law**: Ignoring legal frameworks and judicial processes.

7. **Suppressing Opposition and Dissent**: Marginalizing or persecuting opposition voices and critics.

8. **Lack of Transparency**: Opacity in governmental and institutional operations.

9. **Gender and Social Inequality**: Failing to promote equality and ensure the participation of all societal groups.

1. Corruption and Self-Interest: The Misuse of Public Resources for Personal Gain

Principle: The principle of "Where they tie cow, that is where they eat" encapsulates a deeply ingrained practice of corruption and self-interest, where public officials exploit their positions of power to siphon resources meant for public welfare into their private coffers. This misuse of public resources not only undermines the equitable distribution of national wealth but also erodes public trust in governance, hinders economic development, and perpetuates social inequality.

Context and Importance in Sierra Leone: Corruption remains one of the most significant impediments to Sierra Leone's development.

Despite numerous efforts to curb this menace, corrupt practices continue to plague various sectors of governance. The detrimental impact of corruption in Sierra Leone is multifaceted:

1. **Economic Stagnation**: Corruption diverts resources away from critical areas such as healthcare, education, and infrastructure, resulting in substandard services and stunted economic growth.

2. **Erosion of Public Trust**: When public officials engage in corrupt activities, it undermines the trust that citizens have in their government. This lack of trust can lead to apathy, reduced civic engagement, and an unwillingness to comply with laws and regulations.

3. **Social Inequality**: Corruption disproportionately affects the poorest and most vulnerable populations, who rely most on public services. When resources are diverted, these groups are often left without access to essential services.

4. **Weakening of Institutions**: Persistent corruption undermines the effectiveness and credibility of public institutions, including the judiciary, law enforcement, and regulatory bodies, making it difficult to enforce laws and policies.

SLAM's Role in Addressing Corruption: SLAM plays a critical role in the fight against corruption by advocating for transparency, accountability, and good governance. SLAM's mission includes:

- **Raising Awareness**: Educating the public about the detrimental effects of corruption and the importance of integrity in public service.

- **Promoting Transparency**: Encouraging the adoption of transparent practices in government operations to reduce opportunities for corrupt behavior.

- **Advocating for Stronger Legal Frameworks**: Pushing for robust anti-corruption laws and their stringent enforcement.

- **Engaging Citizens**: Empowering citizens to hold their leaders accountable through active participation and oversight.

Strategies for Combating Corruption

1. **Strengthening Legal Frameworks**:

    - Advocate for the development and enforcement of comprehensive anti-corruption laws that include severe penalties for corrupt practices.

    - Promote the establishment of independent anti-corruption bodies with the authority and resources to investigate and prosecute corruption cases.

2. **Public Monitoring Initiatives**:
   - Organize and support initiatives that monitor government projects and expenditures to ensure funds are used appropriately.
   - Train local NGOs and community groups to observe and report on government activities.

3. **Transparency Platforms**:
   - Develop digital platforms that provide the public with access to information about government activities, including budgets, expenditures, and project outcomes.
   - Encourage the government to publish regular reports on its financial activities and audits.

4. **Educational Programs**:
   - Implement programs that educate citizens on the impact of corruption and their role in combating it.
   - Conduct workshops and seminars for public officials on ethical behavior and the importance of transparency.

5. **Partnerships with Anti-Corruption Bodies**:
   - Collaborate with international and local organizations, such as Transparency International, to adopt best practices and global standards in the fight against corruption.
   - Participate in global anti-corruption initiatives to share knowledge and strategies.

6. **Encouraging Whistleblower Protections:**
   - Advocate for the implementation of robust whistleblower protection laws that safeguard individuals who report corrupt activities.
   - Establish secure channels for reporting corruption, ensuring confidentiality and protection against retaliation.

7. **Promoting a Culture of Accountability:**
   - Foster a culture of accountability within public institutions by ensuring that officials are held responsible for their actions.
   - Support the development of performance-based accountability systems for public officials and agencies.

8. **Engagement in Policy Discussions:**
   - Actively participate in policy discussions and public forums where governance issues are debated.
   - Use these platforms to influence policy decisions and advocate for measures that enhance accountability and reduce corruption.

Constitution Reference

- **Fundamental Principles of State Policy** (Chapter II, Section 5): Calls for the eradication of corrupt practices and the abuse of power.

- **Duties of the Citizen** (Chapter II, Section 13): Encourages citizens to prevent the misappropriation and squandering of public funds.

Book Reference: "Corruption and Government: Causes, Consequences, and Reform" by Susan Rose-Ackerman and Bonnie J. Palifka. This book provides a comprehensive analysis of the causes and consequences of corruption, as well as strategies for reform.[15] It examines how corruption undermines governance and development, offering insights into effective anti-corruption measures. The insights from this book will guide SLAM's advocacy efforts in Sierra Leone.[16]

Conclusion: Addressing the deeply rooted issue of corruption and self-interest is crucial for Sierra Leone's progress. By implementing these strategies and fostering a culture of integrity, SLAM can significantly contribute to the reduction of corruption and the promotion of transparent, accountable governance. This, in turn, will help build a more equitable and prosperous society where resources are used for the benefit of all citizens, in line with the principles outlined in the 1991 Constitution of Sierra Leone.

---

[15] "Corruption and Government: Causes, Consequences, and Reform" (by Susan Rose-Ackerman and Bonnie J. Palifka).

[16] Corruption and Government: Causes, Consequences, and Reform

2. Nepotism and Favoritism: Prioritizing Family, Friends, or Close Associates Over Qualified Individuals

Principle: Nepotism and favoritism refer to the practice of giving preferential treatment to family members, friends, or close associates, often disregarding merit and qualifications. This principle undermines the fairness and integrity of governance, leading to inefficiency, lack of trust, and unequal opportunities.

Context and Importance in Sierra Leone: Nepotism and favoritism are deeply ingrained issues in Sierra Leone, as in many other countries. These practices have significant negative impacts on governance and development:

1. **Erosion of Meritocracy**: Favoritism erodes the principle of meritocracy, where positions and opportunities should be awarded based on skills, experience, and qualifications.

2. **Decreased Efficiency**: Appointing less qualified individuals to key positions reduces the overall efficiency and effectiveness of public institutions and services.

3. **Undermined Public Trust**: When citizens perceive that success is determined by connections rather than merit, it erodes trust in public institutions and discourages public engagement and participation.

4. **Perpetuation of Inequality**: Favoritism perpetuates social inequality, as those without connections are systematically disadvantaged, leading to a lack of social mobility and increased frustration among the populace.

SLAM's Role in Addressing Nepotism and Favoritism: SLAM can play a crucial role in combating nepotism and favoritism by advocating for merit-based practices and promoting fairness and equality in governance. SLAM's initiatives can include:

- **Raising Awareness**: Educating the public and public officials about the negative impacts of nepotism and the importance of merit-based systems.

- **Advocating for Policies**: Pushing for the implementation and enforcement of policies that promote transparency and meritocracy in public appointments and resource allocation.

- **Monitoring and Reporting**: Establishing mechanisms to monitor and report instances of nepotism and favoritism in public institutions.

Strategies for Combating Nepotism and Favoritism

1. **Strengthening Merit-Based Recruitment and Promotion**:
    - Advocate for clear and transparent criteria for recruitment and promotion in public institutions.
    - Support the establishment of independent recruitment bodies that ensure positions are filled based on merit.

2. **Transparency and Accountability Mechanisms**:
    - Promote the publication of recruitment processes and criteria to ensure transparency.

- Encourage the use of independent panels for hiring and promotion decisions to minimize bias.

3. **Educational Programs:**
    - Conduct workshops and training sessions for public officials on the importance of meritocracy and the negative impacts of nepotism and favoritism.
    - Educate citizens on their rights to fair treatment and the mechanisms available to report instances of favoritism.

4. **Policy Advocacy:**
    - Advocate for the implementation of policies that penalize nepotism and favoritism, including stringent anti-nepotism laws.
    - Push for regular audits and reviews of recruitment and promotion processes in public institutions.

5. **Whistleblower Protections:**
    - Advocate for robust whistleblower protection laws to safeguard individuals who report nepotism and favoritism.
    - Establish confidential channels for reporting unfair practices.

Constitution Reference

- **Fundamental Principles of State Policy** (Chapter II, Section 8): Promotes social justice and equality of opportunity, emphasizing the need for a merit-based system.

- **Duties of the Citizen** (Chapter II, Section 13): Encourages citizens to contribute to the advancement of their community, which includes advocating for fair and just governance practices.

Book Reference: "The Meritocracy Trap: How America's Foundational Myth Feeds Inequality, Dismantles the Middle Class, and Devours the Elite" by Daniel Markovits. This book explores the concept of meritocracy and its implications for society, providing a critical examination of how deviations from merit-based practices can lead to widespread inequality and inefficiency.[17] By implementing these strategies and leveraging the insights from established research, SLAM can significantly contribute to reducing nepotism and favoritism in Sierra Leone.[18]

---

[17] "The Meritocracy Trap: How America's Foundational Myth Feeds Inequality, Dismantles the Middle Class, and Devours the Elite" (Daniel Markovits),
[18] The Meritocracy Trap: How America's Foundational Myth Feeds Inequality, Dismantles the Middle Class, and Devours the Elite

3. Regionalism and Ethnocentrism: Focusing on the Well-Being of One's Own Region or Ethnic Group While Neglecting Others

Principle: Regionalism and ethnocentrism involve prioritizing the interests and well-being of one's own region or ethnic group over those of other regions or groups. This principle undermines national unity, fosters division, and hinders equitable development. When public policies and resources are distributed based on regional or ethnic favoritism, it leads to disparities and conflicts that can destabilize the nation.

Context and Importance in Sierra Leone: Sierra Leone is a diverse country with multiple ethnic groups and regions, each with its own unique cultural and historical background. However, regionalism and ethnocentrism have historically contributed to social and political tensions, hampering national unity and development. The detrimental impacts include:

1. **Social Fragmentation**: Prioritizing one group over others fosters division and resentment, weakening the social fabric.

2. **Inequitable Development**: Resources and development efforts concentrated in certain regions lead to significant disparities in infrastructure, healthcare, education, and economic opportunities.

3. **Political Instability**: Ethnocentric and regional biases can fuel political unrest and conflict, undermining the stability necessary for sustained development.

4. **Erosion of National Identity**: When regional and ethnic identities overshadow national identity, it hampers efforts to build a cohesive and unified nation.

SLAM's Role in Addressing Regionalism and Ethnocentrism: SLAM is committed to promoting national unity and equitable development by addressing the issues of regionalism and ethnocentrism. SLAM's initiatives can focus on fostering inclusivity, advocating for fair resource distribution, and educating citizens on the importance of national cohesion.

Strategies for Combating Regionalism and Ethnocentrism

1. **Promoting Inclusive Policies:**

    - **Objective**: Ensure that policies and development programs are inclusive and equitable, benefiting all regions and ethnic groups.

    - **Implementation**:

        - **Equitable Resource Allocation**: Advocate for policies that ensure fair distribution of resources across all regions. This includes funding for infrastructure, education, healthcare, and other public services.

        - **Inclusive Development Projects**: Promote development projects that consider the needs of all regions and communities, ensuring balanced growth and opportunities.

2. **Fostering National Unity:**

    - **Objective**: Strengthen national identity and unity through inclusive and participatory initiatives.

- **Implementation:**
  - **Cultural Exchange Programs:** Organize cultural exchange programs and events that celebrate the diversity of Sierra Leone's ethnic groups, fostering mutual understanding and respect.
  - **National Service Programs:** Encourage participation in national service programs that bring together individuals from different regions and ethnic backgrounds to work on common goals.

3. **Educational Campaigns:**
   - **Objective:** Educate citizens on the dangers of regionalism and ethnocentrism and the importance of national unity.
   - **Implementation:**
     - **Curriculum Reform:** Advocate for the inclusion of national unity and diversity education in school curricula to instill values of inclusivity from a young age.
     - **Public Awareness Campaigns:** Conduct public awareness campaigns through media, workshops, and community meetings to highlight the benefits of unity and the dangers of divisiveness.

4. **Encouraging Political Representation:**
   - **Objective:** Ensure diverse and balanced representation in political and governance structures.
   - **Implementation:**
     - **Proportional Representation:** Advocate for electoral systems and policies that ensure proportional representation of all regions and ethnic groups in government.
     - **Diverse Leadership:** Promote the inclusion of leaders from various regions and ethnic backgrounds in key decision-making positions.

5. **Monitoring and Reporting:**
   - **Objective:** Monitor and report instances of regional and ethnic favoritism in policy-making and resource distribution.
   - **Implementation:**
     - **Watchdog Initiatives:** Establish watchdog groups to oversee government activities and ensure equitable treatment of all regions and ethnic groups.
     - **Transparency Reports:** Publish regular reports on resource allocation and development efforts, highlighting any disparities and advocating for corrective actions.

Constitution Reference

- **Fundamental Principles of State Policy** (Chapter II, Section 5): Promotes national integration and unity and discourages discrimination on grounds of place of origin, ethnicity, or other affiliations.

- **Duties of the Citizen** (Chapter II, Section 13): Encourages citizens to cultivate a sense of nationalism and patriotism, placing loyalty to the state above sectional loyalties.

Book References: 1. "Ethnicity and the Politics of Democratic Nation-Building in Africa" by Bruce Berman, Dickson Eyoh, and Will Kymlicka. This book explores the challenges and opportunities of nation-building in ethnically diverse African countries. It provides insights into how ethnic diversity can be managed to foster national unity and democratic governance.[19]

2. **"Ethnic Conflict and Civic Life: Hindus and Muslims in India"** by Ashutosh Varshney. This book explores the dynamics of ethnic conflict and the role of civic engagement in fostering interethnic cooperation. It provides valuable insights into the importance of building inclusive societies and the dangers of ethnic division. The insights from this book could guide SLAM's efforts to combat regionalism and ethnocentrism in Sierra Leone.[20]

---

[19] Ethnicity and the Politics of Democratic Nation-Building in Africa.
[20] Ethnic Conflict and Civic Life: Hindus and Muslims in India.

4. Political Partisanship: Placing Party Loyalty Above National Interest

Principle: Political partisanship involves placing the interests and loyalty of one's political party above the broader national interest. This principle undermines the democratic process, as it prioritizes party agendas over the common good, often leading to gridlock, inefficiency, and polarized governance. When political decisions are driven by party loyalty rather than the national interest, it erodes public trust in government institutions and hampers effective governance.

Context and Importance in Sierra Leone: In Sierra Leone, political partisanship has historically played a significant role in shaping the political landscape. The detrimental impacts of excessive partisanship include:

1. **Governance Gridlock**: When political leaders prioritize party loyalty, it can lead to stalemates in decision-making processes, delaying critical policies and reforms.

2. **Policy Instability**: Partisan politics can result in frequent policy reversals, as successive governments undo the work of their predecessors, leading to a lack of continuity in governance.

3. **Public Disillusionment**: Citizens become disillusioned with the political process when they perceive that leaders are more focused on party interests than on addressing national challenges.

4. **Exclusion of Minority Voices**: Partisan politics can marginalize minority groups and opposition voices, reducing the inclusiveness and representativeness of governance.

SLAM's Role in Addressing Political Partisanship: SLAM can play a pivotal role in mitigating the negative effects of political partisanship by promoting national unity, fostering dialogue, and advocating for policies that prioritize the national interest over party agendas. SLAM's initiatives can focus on encouraging bipartisan cooperation, enhancing civic education, and monitoring political practices.

Strategies for Combating Political Partisanship

1. **Promoting Bipartisan Cooperation:**
    - **Objective**: Encourage political parties to work together on issues of national importance.
    - **Implementation**:
        - **National Unity Forums**: Organize forums and roundtable discussions that bring together leaders from different political parties to discuss and collaborate on critical national issues.
        - **Joint Policy Committees**: Advocate for the creation of joint policy committees composed of members from multiple parties to address key challenges such as healthcare, education, and economic development.

2. **Civic Education Programs:**
    - **Objective**: Educate citizens on the importance of placing national interest above party loyalty.

- Implementation:
  - **Educational Campaigns**: Conduct public awareness campaigns that highlight the negative impacts of excessive partisanship and the benefits of bipartisan cooperation.
  - **School Curricula**: Advocate for the inclusion of civic education in school curricula, emphasizing the importance of national unity and the dangers of political polarization.

3. **Monitoring Political Practices**:
   - **Objective**: Monitor and report on political practices that prioritize party loyalty over the national interest.
   - Implementation:
     - **Political Watchdog Groups**: Establish watchdog groups to oversee political activities and report instances of excessive partisanship.
     - **Transparency Reports**: Publish regular reports on the alignment of political decisions with national interests, highlighting both positive and negative examples.

4. **Encouraging Inclusive Governance**:
    - **Objective**: Promote governance structures that include voices from all political spectrums.
    - **Implementation**:
        - **Inclusive Decision-Making Bodies**: Advocate for the inclusion of opposition members in key decision-making bodies to ensure diverse perspectives are considered.
        - **Proportional Representation**: Support electoral systems that encourage proportional representation, giving minority parties a voice in governance.

5. **Building a Culture of National Interest**:
    - **Objective**: Foster a political culture where the national interest is prioritized over party loyalty.
    - **Implementation**:
        - **National Interest Pledge**: Encourage political leaders to publicly commit to prioritizing the national interest through a formal pledge.
        - **Recognition of Bipartisan Efforts**: Recognize and reward instances of successful bipartisan cooperation in addressing national challenges.

Constitution Reference

- **Fundamental Principles of State Policy** (Chapter II, Section 5): Emphasizes the importance of national unity and the need for governance that prioritizes the welfare of the entire population.

- **Duties of the Citizen** (Chapter II, Section 13): Encourages citizens to cultivate a sense of nationalism and patriotism, placing loyalty to the state above sectional loyalties.

Book Reference: "Political Parties in Africa: Challenges for Sustained Multiparty Democracy." This book examines the role of political parties in Africa and the challenges they face in promoting sustained multiparty democracy.[21] It provides insights into how political partisanship can be managed to foster national unity and effective governance. The insights from this book can guide SLAM's efforts to address political partisanship and promote national interest in Sierra Leone.[22]

5. Patronage Politics: Distributing Public Resources and Favors in Exchange for Political Support

Principle: Patronage politics involves distributing public resources and favors in exchange for political support. This practice undermines democratic principles by prioritizing loyalty and political gain over merit and public interest.

---

[21] "Political Parties in Africa: Challenges for Sustained Multiparty Democracy
[22] Challenges for Sustained Multiparty Democracy.

Patronage can lead to inefficient use of resources, corruption, and a lack of accountability in governance, as decisions are made based on political considerations rather than the needs of the populace.

Context and Importance in Sierra Leone: In Sierra Leone, patronage politics has been a pervasive issue, affecting various aspects of governance and development. The detrimental impacts of patronage politics include:

1. **Inefficiency and Misallocation of Resources**: Resources are often allocated based on political loyalty rather than need, leading to inefficiencies and mismanagement.

2. **Corruption**: Patronage systems are prone to corruption, as public officials may engage in bribery and favoritism to secure political support.

3. **Erosion of Public Trust**: When citizens perceive that resources and opportunities are distributed based on political connections, it undermines trust in public institutions.

4. **Weakening of Institutions**: Patronage politics can weaken democratic institutions by promoting individuals based on loyalty rather than competence, reducing the effectiveness of public services.

SLAM's Role in Addressing Patronage Politics: SLAM can play a critical role in combating patronage politics by advocating for transparent and merit-based systems, promoting accountability, and raising public awareness. SLAM's initiatives can focus on fostering a culture of integrity and advocating for reforms that reduce the influence of patronage in governance.

Strategies for Combating Patronage Politics

1. **Promoting Merit-Based Systems**:
   - **Objective**: Ensure that public resources and opportunities are distributed based on merit rather than political loyalty.
   - **Implementation**:
     - **Transparent Recruitment Processes**: Advocate for transparent and merit-based recruitment processes in public institutions.
     - **Independent Commissions**: Support the establishment of independent commissions to oversee public appointments and resource allocation.

2. **Enhancing Accountability and Transparency**:
   - **Objective**: Promote accountability and transparency in the distribution of public resources.
   - **Implementation**:
     - **Public Disclosure of Allocations**: Advocate for the public disclosure of resource allocations and government contracts to ensure transparency.
     - **Audit and Oversight Mechanisms**: Support the implementation of regular audits and oversight mechanisms to monitor the use of public resources.

3. **Public Awareness Campaigns:**
    - **Objective**: Educate citizens about the negative impacts of patronage politics and the importance of merit-based governance.
    - **Implementation**:
        - **Educational Workshops**: Conduct workshops and seminars to educate the public on the importance of transparency and accountability in governance.
        - **Media Campaigns**: Use media campaigns to highlight instances of patronage and promote a culture of integrity.

4. **Advocating for Legal and Policy Reforms:**
    - **Objective**: Advocate for legal and policy reforms that reduce the influence of patronage in governance.
    - **Implementation**:
        - **Anti-Corruption Legislation**: Advocate for the enactment and enforcement of anti-corruption legislation that targets patronage practices.
        - **Policy Frameworks**: Support the development of policy frameworks that promote merit-based systems and reduce political interference in public appointments.

5. **Supporting Civil Society Organizations**:
    - **Objective**: Empower civil society organizations to monitor and report on patronage practices.
    - **Implementation**:
        - **Capacity Building**: Provide training and resources to civil society organizations to enhance their capacity to monitor and report on governance issues.
        - **Collaborative Networks**: Foster collaborative networks among civil society organizations to share best practices and coordinate efforts to combat patronage.

Constitution Reference

- **Fundamental Principles of State Policy** (Chapter II, Section 5): Emphasizes the need for governance based on the principles of transparency, accountability, and meritocracy.

- **Duties of the Citizen** (Chapter II, Section 13): Encourages citizens to resist corruption and the misuse of public resources.

Book Reference: "The African State: Reconsiderations." This book provides an in-depth analysis of the nature of the African state, including the challenges of patronage politics. It offers insights into how patronage systems operate and the impact they have on governance and development in Africa.

The insights from this book could guide SLAM's efforts to address patronage and promote good governance in Sierra Leone.[23]

6. Disregard for the Rule of Law: Ignoring Legal Frameworks and Judicial Processes

Principle: Disregard for the rule of law involves ignoring established legal frameworks and judicial processes, leading to arbitrary governance and a lack of accountability. This principle undermines democracy and governance by creating an environment where laws are applied inconsistently and where individuals and institutions operate above the law. Respect for the rule of law is essential for maintaining order, protecting human rights, and ensuring fair and predictable governance.

Context and Importance in Sierra Leone: In Sierra Leone, the disregard for the rule of law has been a significant challenge, contributing to various governance issues. The detrimental impacts include:

1. **Erosion of Trust in Institutions**: When laws are not consistently applied, public trust in governmental and judicial institutions erodes.

2. **Increased Corruption**: Disregard for the rule of law facilitates corruption, as individuals and institutions exploit the lack of accountability for personal gain.

---

[23] The African State: Reconsiderations (link).

3. **Human Rights Violations**: Ignoring legal frameworks often leads to the violation of citizens' rights, as there are no mechanisms to protect them.

4. **Political Instability**: Arbitrary governance can lead to political unrest and instability as citizens and opposition groups react against perceived injustices.

SLAM's Role in Addressing Disregard for the Rule of Law: SLAM can play a critical role in promoting the rule of law by advocating for the consistent application of legal frameworks, supporting judicial independence, and raising public awareness about the importance of the rule of law. SLAM's initiatives can focus on strengthening legal institutions, enhancing transparency, and ensuring accountability.

Strategies for Combating Disregard for the Rule of Law

1. **Strengthening Judicial Independence**:
    - **Objective**: Ensure that the judiciary operates independently and is free from political interference.
    - **Implementation**:
        - **Judicial Reforms**: Advocate for judicial reforms that protect judges from political pressure and ensure fair appointments based on merit.
        - **Support for Judicial Bodies**: Provide support for judicial bodies to enhance their capacity to operate independently and efficiently.

2. **Promoting Legal Literacy and Awareness**:
   - **Objective**: Educate citizens about their legal rights and the importance of the rule of law.
   - **Implementation**:
     - **Public Education Campaigns**: Conduct public education campaigns through media, workshops, and community meetings to raise awareness about legal rights and judicial processes.
     - **School Curricula**: Advocate for the inclusion of legal education in school curricula to instill respect for the rule of law from a young age.

3. **Enhancing Transparency and Accountability**:
   - **Objective**: Promote transparency and accountability in the application of laws and judicial processes.
   - **Implementation**:
     - **Public Access to Legal Information**: Advocate for public access to legal information, including laws, regulations, and court decisions.
     - **Monitoring and Reporting**: Establish mechanisms to monitor and report instances of legal violations and judicial misconduct.

4. **Supporting Anti-Corruption Measures**:
   - **Objective**: Combat corruption by ensuring the rule of law is upheld.
   - **Implementation**:
     - **Anti-Corruption Legislation**: Advocate for the enactment and enforcement of strong anti-corruption laws.
     - **Whistleblower Protections**: Support the implementation of whistleblower protection laws to encourage reporting of legal violations and corruption.

5. **Building Capacity of Legal Institutions**:
   - **Objective**: Strengthen the capacity of legal institutions to enforce laws effectively.
   - **Implementation**:
     - **Training Programs**: Organize training programs for judges, lawyers, and law enforcement officials on best practices in legal and judicial processes.
     - **Resource Allocation**: Advocate for adequate resources to be allocated to legal institutions to enhance their operational capacity.

Constitution Reference

- **Recognition and Protection of Fundamental Human Rights and Freedoms** (Chapter III, Sections 15-28): Emphasizes the protection of human rights through the rule of law.

- **Judicial Independence** (Chapter VII, Sections 120-122): Stipulates the independence of the judiciary as a fundamental principle of governance.

Book Reference: "The Rule of Law in the 21st Century: A Worldwide Perspective". The rule of law is sometimes expressed as 'no person is above the law.' A more comprehensive description of the concept has been elusive for generations of scholars, lawyers, and judges. What does the phrase mean? More specifically, what does the rule of law mean in the context of 21st-century issues and challenges? It explores the factors that undermine the rule of law and offers strategies for strengthening legal frameworks and judicial processes.[24] The insights from this book could guide SLAM's efforts to promote legal frameworks and judicial integrity in Sierra Leone.

7. Suppressing Opposition and Dissent: Marginalizing or Persecuting Opposition Voices and Critics

Principle: Suppressing opposition and dissent involves marginalizing, persecuting, or silencing individuals or groups who express opposing viewpoints or criticize the government.

---

[24] The Rule of Law in the 21st Century: A Worldwide Perspective.

This principle is antithetical to democratic values, as it undermines the free exchange of ideas, stifles political pluralism, and inhibits accountability. A healthy democracy relies on the ability of all citizens to voice their opinions and participate in governance without fear of retribution.

Context and Importance in Sierra Leone: In Sierra Leone, the suppression of opposition and dissent has been a significant issue, contributing to political instability and weakening democratic governance. The detrimental impacts include:

1. **Erosion of Democratic Principles**: Suppressing dissent undermines fundamental democratic principles such as freedom of speech, assembly, and association.

2. **Lack of Accountability**: When opposition voices are silenced, it reduces government accountability and transparency, as there are fewer checks on power.

3. **Political Instability**: Persecution of opposition figures can lead to political unrest and conflict, destabilizing the nation.

4. **Human Rights Violations**: Suppressing dissent often involves human rights abuses, including arbitrary detention, harassment, and violence against critics.

SLAM's Role in Addressing Suppression of Opposition and Dissent: SLAM can play a crucial role in promoting political pluralism and protecting the rights of opposition voices and critics. SLAM's initiatives can focus on advocating for legal protections, supporting civil society organizations, and raising public awareness about the importance of political diversity and freedom of expression.

Strategies for Combating Suppression of Opposition and Dissent

1. **Advocating for Legal Protections**:
   - **Objective**: Ensure that laws protect the rights of opposition voices and critics.
   - **Implementation**:
     - **Freedom of Speech and Assembly Laws**: Advocate for robust legal protections for freedom of speech, assembly, and association.
     - **Judicial Independence**: Support judicial independence to ensure that courts can act impartially in cases involving political dissent.

2. **Supporting Civil Society Organizations**:
   - **Objective**: Empower civil society organizations to defend and promote political pluralism.
   - **Implementation**:
     - **Capacity Building**: Provide training and resources to civil society organizations to enhance their capacity to monitor and report on human rights abuses and political repression.
     - **Advocacy Networks**: Foster networks of civil society organizations to coordinate advocacy efforts and share best practices.

3. **Promoting Media Freedom and Pluralism**:
    - **Objective**: Ensure that media outlets can operate freely and provide diverse perspectives.
    - **Implementation**:
        - **Legal Protections for Journalists**: Advocate for laws that protect journalists from harassment and violence.
        - **Support for Independent Media**: Provide support for independent media outlets to ensure that they can operate without government interference.

4. **Raising Public Awareness**:
    - **Objective**: Educate the public about the importance of political diversity and the dangers of suppressing dissent.
    - **Implementation**:
        - **Public Education Campaigns**: Conduct public education campaigns through media, workshops, and community meetings to highlight the importance of opposition voices in a democracy.
        - **School Curricula**: Advocate for the inclusion of civic education in school curricula, emphasizing the value of political pluralism and free expression.

5. **Monitoring and Reporting**:

    o **Objective**: Monitor and report instances of political repression and human rights abuses.

    o **Implementation**:

    - **Human Rights Watchdog Groups**: Establish watchdog groups to monitor and report on the treatment of opposition voices and critics.

    - **Transparency Reports**: Publish regular reports on political repression, highlighting abuses and advocating for reforms.

Constitution Reference

- **Recognition and Protection of Fundamental Human Rights and Freedoms** (Chapter III, Sections 15-28): Emphasizes the protection of fundamental rights, including freedom of speech, assembly, and association.

- **Political Rights** (Chapter IV, Sections 31-32): Guarantees the right to participate in political activities and to form or join political parties.

Book Reference: "Democracy and Human Rights in Africa". This book provides an in-depth examination of the state of democracy and human rights in Africa. It explores the challenges faced by opposition groups and the importance of protecting political pluralism and freedom of expression.[25]

---

[25] Democracy and Human Rights in Africa

The insights from this book could guide SLAM's efforts to protect opposition voices and promote democratic governance in Sierra Leone.

8. Lack of Transparency: Opacity in Governmental and Institutional Operations

Principle: A lack of transparency in governmental and institutional operations refers to the absence of openness and accessibility in decision-making processes, financial management, and public service delivery. Opacity undermines public trust, enables corruption, and impedes accountability. Transparency is essential for ensuring that governments operate in the best interests of their citizens and that public officials are held accountable for their actions.

Context and Importance in Sierra Leone: In Sierra Leone, opacity in government operations has been a significant barrier to effective governance and development. The detrimental impacts of a lack of transparency include:

1. **Increased Corruption**: Opacity creates opportunities for corrupt practices, as it becomes easier for officials to misuse public resources without detection.

2. **Erosion of Public Trust**: When government operations are opaque, citizens lose trust in their leaders and institutions, leading to disengagement and apathy.

3. **Inefficiency and Mismanagement**: Without transparency, it is difficult to ensure that resources are allocated and used efficiently, leading to waste and mismanagement.

4. **Impaired Accountability**: Opacity makes it challenging to hold public officials accountable for their actions, reducing the effectiveness of oversight mechanisms.

SLAM's Role in Addressing Lack of Transparency: SLAM can play a pivotal role in promoting transparency by advocating for open government practices, supporting the development of transparency platforms, and educating citizens about their right to access information. SLAM's initiatives can focus on enhancing transparency in financial management, decision-making processes, and public service delivery.

Strategies for Combating Lack of Transparency

1. **Advocating for Open Government Practices**:
    - **Objective**: Ensure that government operations are conducted openly and transparently.
    - **Implementation**:
        - **Freedom of Information Laws**: Advocate for the enactment and enforcement of robust freedom of information laws that guarantee public access to government records and data.
        - **Transparent Decision-Making**: Promote the establishment of transparent decision-making processes that include public consultations and stakeholder engagement.

2. **Developing Transparency Platforms**:
    - **Objective**: Create digital platforms that provide public access to information about government activities, including budgets, expenditures, and project outcomes.
    - **Implementation**:
        - **Government Websites**: Develop and maintain government websites that publish comprehensive information on financial activities, project statuses, and policy decisions.
        - **Open Data Portals**: Establish open data portals that host datasets on various aspects of governance, such as health, education, infrastructure, and public spending.

3. **Regular Publication of Reports**:
    - **Objective**: Ensure that the government publishes regular and detailed reports on its financial activities and audits.
    - **Implementation**:
        - **Annual Reports**: Advocate for the publication of annual financial reports that detail the government's income, expenditures, and budgetary allocations.

- **Audit Reports**: Promote the regular release of audit reports conducted by independent bodies, highlighting any discrepancies or issues in financial management.

- **Project Updates**: Encourage the government to provide regular updates on the progress and outcomes of major public projects and initiatives.

4. **Public Awareness Campaigns**:

    - **Objective**: Educate citizens about the importance of transparency and their right to access information.

    - **Implementation**:

        - **Educational Workshops**: Conduct workshops and seminars to educate the public on how to access and interpret government data.

        - **Media Campaigns**: Use media campaigns to raise awareness about transparency initiatives and encourage active citizen participation.

5. **Monitoring and Reporting**:

    - **Objective**: Monitor and report instances of opacity and lack of transparency in government operations.

- Implementation:
  - **Transparency Watchdog Groups**: Establish watchdog groups to oversee government activities and ensure compliance with transparency standards.
  - **Transparency Reports**: Publish regular reports on transparency in government operations, highlighting both successes and areas needing improvement.

Constitution Reference

- **Fundamental Principles of State Policy** (Chapter II, Section 8): Emphasizes the need for transparency and accountability in governance.

- **Recognition and Protection of Fundamental Human Rights and Freedoms** (Chapter III, Sections 15-28): Ensures the right of citizens to access information.

Book Reference: "Accountable Government in Africa: Perspectives from Public". The book brings together a number of leading experts in the fields of public law, political science, and democratization studies to discuss problems of accountability, identify ways of making African governments accountable, and describe the extent to which these mechanisms work in practice.[26]

---

[26] Accountable Government in Africa: Perspectives from Public.

9. Gender and Social Inequality: Failing to Promote Equality and Ensure the Participation of All Societal Groups

Principle: Gender and social inequality involve the unequal treatment or perceptions of individuals based on their gender, social status, or other attributes. This principle undermines democratic values by limiting opportunities for certain groups, marginalizing their voices, and perpetuating systemic discrimination. Promoting equality and ensuring the participation of all societal groups are essential for building an inclusive, fair, and democratic society.

Context and Importance in Sierra Leone: In Sierra Leone, gender and social inequality remain significant barriers to development and democratic governance. The detrimental impacts of these inequalities include:

1. **Limited Economic Opportunities**: Gender and social inequality restrict access to education, employment, and economic resources, particularly for women and marginalized groups.

2. **Underrepresentation in Governance**: Women and marginalized groups are often underrepresented in political and decision-making processes, leading to policies that do not fully address their needs and concerns.

3. **Social Exclusion**: Inequality perpetuates social exclusion, reducing social cohesion and increasing vulnerability to poverty and violence.

4. **Stunted Development**: When large segments of the population are excluded from full participation in society, the country's overall development is hindered.

SLAM's Role in Addressing Gender and Social Inequality: SLAM can play a crucial role in promoting gender and social equality by advocating for inclusive policies, supporting the empowerment of marginalized groups, and raising public awareness about the importance of equality. SLAM's initiatives can focus on enhancing educational opportunities, promoting equal representation, and addressing systemic barriers to equality.

Strategies for Combating Gender and Social Inequality

1. **Advocating for Inclusive Policies**:
   - **Objective**: Ensure that policies and laws promote equality and protect the rights of all societal groups.
   - **Implementation**:
     - **Gender Equality Laws**: Advocate for the enactment and enforcement of laws that promote gender equality and protect women's rights.
     - **Social Protection Policies**: Support the development of social protection policies that address the needs of marginalized groups, including those based on ethnicity, disability, and socioeconomic status.

2. **Enhancing Educational Opportunities**:
   - **Objective**: Ensure equal access to quality education for all societal groups.
   - **Implementation**:
     - **Scholarship Programs**: Advocate for and support scholarship programs aimed at increasing educational opportunities for girls and marginalized groups.
     - **Educational Campaigns**: Conduct campaigns to raise awareness about the importance of education for all and to challenge cultural norms that hinder educational attainment for certain groups.

3. **Promoting Equal Representation**:
   - **Objective**: Increase the representation of women and marginalized groups in political and decision-making processes.
   - **Implementation**:
     - **Quotas and Affirmative Action**: Advocate for the implementation of quotas and affirmative action policies to ensure the representation of women and marginalized groups in governance.

- **Leadership Training**: Provide leadership training and capacity-building programs for women and marginalized individuals to enhance their participation in political and public life.

4. **Addressing Systemic Barriers**:
   - **Objective**: Identify and address systemic barriers that perpetuate inequality.
   - **Implementation**:
     - **Policy Analysis**: Conduct analyses to identify policies and practices that contribute to inequality and advocate for their reform.
     - **Public Awareness Campaigns**: Use media and community outreach to challenge stereotypes and promote positive attitudes towards gender and social equality.

5. **Supporting Empowerment Initiatives**:
   - **Objective**: Empower women and marginalized groups through targeted programs and initiatives.
   - **Implementation**:
     - **Economic Empowerment Programs**: Support initiatives that provide women and marginalized groups with access to microfinance, vocational training, and entrepreneurial opportunities.

- **Health and Social Services**: Advocate for and support programs that improve access to healthcare and social services for all societal groups.

Constitution Reference

- **Recognition and Protection of Fundamental Human Rights and Freedoms** (Chapter III, Sections 15-28): Ensures the protection of fundamental human rights, including the right to equality and non-discrimination.

- **Duties of the Citizen** (Chapter II, Section 13): Encourages citizens to promote harmony and the spirit of common brotherhood, regardless of gender, ethnicity, or social status.

Book Reference: "Women and Development in Africa: How Gender Works". This new edition of Women and Development in Africa incorporates the results of more than a decade of new empirical and theoretical research. Michael Kevane provides a broad overview of the sources of underdevelopment in Africa and the role of gender in economic transactions, as well as a cogent analysis of the gendered realities of such issues as land rights, the control of labor, the marriage market, health care, education, and political representation.[27]

---

[27] Women and Development in Africa: How Gender Works

Reflection and Engagement Questions

This chapter delves into the negative factors that undermine democratic governance in Sierra Leone. It examines issues such as corruption, nepotism, regionalism, political partisanship, and other detrimental practices. By identifying these challenges, the chapter aims to provide a clear understanding of the obstacles to good governance. It emphasizes the importance of recognizing and addressing these harmful principles to build a more transparent, accountable, and inclusive political environment. The insights offered are intended to help readers critically assess current governance practices and inspire actionable steps to combat these issues effectively.

1. **Understanding Corruption:**

    - How does corruption manifest in Sierra Leone's governance structures, and what are its most significant impacts on society?

    - _____
      _____

2. **Combatting Nepotism:**

    - What strategies can be implemented to reduce nepotism and promote merit-based appointments in Sierra Leone's public sector?

    - _____
      _____

3. **Addressing Regionalism:**
    - In what ways does regionalism affect political and social cohesion in Sierra Leone, and how can it be mitigated to foster national unity?

    - _____
      _____

4. **Political Partisanship:**
    - How does political partisanship hinder effective governance in Sierra Leone, and what measures can be taken to encourage bipartisan or non-partisan approaches to policy-making?

    - _____
      _____

5. **Promoting Transparency:**
    - Reflect on the current levels of transparency in Sierra Leone's government operations. What steps can be taken to enhance transparency and ensure public access to information?

    - _____
      _____

These questions are designed to help readers engage deeply with the chapter's content, promoting critical thinking and encouraging them to consider practical solutions to the challenges identified.

CHAPTER 9:
STRATEGIES FOR COMBATING DETRIMENTAL PRINCIPLES

Introduction: Implementing good governance principles across Sierra Leone's government institutions is essential for achieving a transparent, accountable, and participatory governance system. SLAM focuses on advocacy and strategic initiatives that will significantly contribute to Sierra Leone's political, social, and economic development, fostering a democratic society where every citizen can thrive.

This report outlines the crucial implementation principles of good governance and democratic values necessary for Sierra Leone's development. It emphasizes accountability, transparency, responsiveness, consensus, equity, effectiveness, rule of law, and participation. These principles serve as a framework for SLAM to enhance governance through advocacy and engagement.

Department and Institutional Application

1. Ministry of Justice

- **Focus**: Strengthen legal frameworks to ensure fairness and justice.

Establish Robust Legal Frameworks to Enforce Accountability and Transparency in Judicial Proceedings

- **Draft and Enact Comprehensive Legal Reforms:**
  - **Actions:**
    - Collaborate with legal experts, civil society organizations, and international bodies to draft and propose new laws or amendments to existing laws that enhance judicial accountability and transparency.
    - Conduct public consultations to gather input from various stakeholders, including the legal community, NGOs, and the general public, to ensure the reforms address the needs and concerns of all parties.
    - Work with Parliament to pass these reforms, ensuring they are in line with international standards and best practices.
  - **Expected Outcomes:**
    - Strengthened legal frameworks that provide clear guidelines and standards for judicial conduct and procedures.

- Increased public trust in the legal system due to more transparent and accountable judicial processes.

- **Set Up an Independent Oversight Body:**
  - **Actions:**
    - Establish an independent body, such as a Judicial Ombudsman or a Judicial Service Commission, with the authority to monitor and evaluate the performance of judges and other judicial officers.
    - Ensure this body has the power to investigate complaints, conduct disciplinary hearings, and enforce sanctions against judicial officers found guilty of misconduct.
    - Provide the oversight body with adequate resources and training to carry out its functions effectively.
  - **Expected Outcomes:**
    - Improved oversight and accountability within the judiciary, leading to a reduction in judicial misconduct and corruption.
    - Enhanced public confidence in the integrity and impartiality of the judiciary.

- **Promote Open Courts:**
  - **Actions:**
    - Implement policies that mandate court proceedings to be open to the public and media, with exceptions only for cases involving sensitive information (e.g., national security or privacy concerns).
    - Develop guidelines for live-streaming significant court cases and publishing court decisions and records online.
  - **Expected Outcomes:**
    - Greater transparency in judicial proceedings, allowing the public and media to scrutinize court processes and decisions.
    - Increased public understanding and trust in the judicial system.
- **Technology Integration:**
  - **Actions:**
    - Develop and implement a comprehensive judicial management information system (JMIS) to track case progress, monitor judicial performance, and manage court schedules.

- Ensure the JMIS is user-friendly and accessible to the public, allowing individuals to check the status of their cases and access court rulings and decisions online.
- **Expected Outcomes**:
  - Improved efficiency in case management and judicial administration.
  - Enhanced transparency and accessibility of judicial information to the public.
- **Training and Capacity Building**:
  - **Actions**:
    - Conduct regular training programs for judges, magistrates, and other judicial staff on topics such as judicial ethics, accountability, transparency, and the importance of public trust in the judiciary.
    - Partner with international legal organizations and universities to provide advanced training and exchange programs for judicial officers.

- **Expected Outcomes**:
    - Enhanced knowledge and skills among judicial officers, leading to better adherence to ethical standards and judicial accountability.
    - A more competent and professional judiciary that is better equipped to handle complex legal issues.

Conduct Regular Audits and Public Reporting to Uphold the Rule of Law

- **Regular Financial Audits**:
    - **Actions**:
        - Implement a system of regular financial audits for all judicial institutions conducted by independent audit firms to ensure impartiality.
        - Publish the results of these audits in annual reports that are accessible to the public and stakeholders.
    - **Expected Outcomes**:
        - Greater financial transparency and accountability within the judiciary.
        - Improved public trust in the

management and use of judicial resources.

- **Performance Audits:**
  - **Actions:**
    - Conduct performance audits to evaluate the efficiency and effectiveness of judicial processes, focusing on areas such as case backlog, timeliness of rulings, and adherence to procedural standards.
    - Use the findings from these audits to identify areas for improvement and implement necessary reforms.
  - **Expected Outcomes:**
    - Enhanced efficiency and effectiveness of the judiciary.
    - Reduced case backlog and faster resolution of cases.
- **Public Reporting:**
  - **Actions:**
    - Mandate the publication of audit reports, judicial performance data, and other relevant information on government websites and in public records.
    - Develop clear and accessible

formats for these reports to ensure they are understandable to the general public.

- **Expected Outcomes**:
    - Increased transparency and public scrutiny of judicial activities.
    - Enhanced public knowledge and awareness of the functioning and performance of the judiciary.

- **Feedback Mechanisms**:
    - **Actions**:
        - Establish mechanisms for public feedback on judicial services, such as suggestion boxes in courts, online feedback forms, and public forums.
        - Regularly review and analyze the feedback received to identify issues and areas for improvement.
    - **Expected Outcomes**:
        - Improved responsiveness of the judiciary to public concerns and needs.
        - Continuous improvement in the quality of judicial services based on public input.

Conclusion: By implementing these detailed actions, the Ministry of Justice can significantly strengthen legal frameworks, enhance judicial accountability and transparency, and uphold the rule of law in Sierra Leone. Establishing robust legal frameworks, setting up independent oversight bodies, promoting open courts, integrating technology, and conducting regular training will ensure a more transparent and accountable judiciary. Regular financial and performance audits, coupled with public reporting and feedback mechanisms, will further enhance public trust. These efforts will contribute to building a fair, just, and democratic society where every citizen can trust and rely on the judicial system.

2. Ministry of Internal Affairs

- **Focus**: Implement training programs for police and security forces on human rights.

Implementation:

Develop Comprehensive Training Programs

- Actions:
    - **Needs Assessment**: Conduct a thorough needs assessment to identify the specific human rights training needs of police and security forces. This should involve consultations with human rights organizations, community leaders, and the police force itself.

- **Curriculum Development**: Develop a detailed training curriculum that covers human rights principles, ethical policing, community engagement, and the importance of transparency and accountability in law enforcement. Include case studies, role-playing scenarios, and interactive sessions to enhance understanding and retention.

- **Training Modules**: Create specific training modules focusing on:

  - International Human Rights Standards: Educate on global human rights frameworks, such as the Universal Declaration of Human Rights and relevant UN conventions.

  - National Human Rights Laws: Ensure police are well-versed in Sierra Leone's own legal commitments to human rights.

  - Ethical Policing Practices: Emphasize the importance of integrity, impartiality, and respect in policing.

  - Community Policing: Teach strategies for building trust and collaboration with communities.

- - **Accountability Mechanisms:** Highlight the procedures for accountability and the consequences of human rights violations.

  - **Trainer Selection and Training:** Select and train a cadre of trainers from within the police force and human rights organizations to deliver these modules effectively.

- **Expected Outcomes:**

  - Increased awareness and understanding of human rights among police and security forces.

  - Improved behavior and attitudes of law enforcement personnel towards citizens.

  - Enhanced trust and cooperation between the police and the communities they serve.

Regular Assessments and Evaluations

- Actions:

  - **Assessment Framework:** Develop a framework for regular assessments of police and security forces' adherence to human rights standards. This should include both quantitative and qualitative measures.

- **Periodic Evaluations**: Conduct periodic evaluations through surveys, interviews, and field observations to assess the effectiveness of the training programs and the extent to which human rights are being respected in practice.

- **Feedback Mechanisms**: Establish feedback mechanisms that allow community members to report on their interactions with police and provide input on the human rights performance of law enforcement personnel.

- **Continuous Improvement**: Use the data collected from assessments and evaluations to continuously improve the training programs. Update the curriculum and training methods based on feedback and identified gaps.

- **Expected Outcomes**:
  - Ongoing improvement in the adherence of police and security forces to human rights standards.
  - Enhanced effectiveness of human rights training programs.
  - Increased accountability and responsiveness of law enforcement agencies.

Establish a Monitoring and Reporting System

- Actions:

    - **Independent Monitoring Body**: Establish an independent body to monitor police activities and investigate human rights violations. This body should have the authority to conduct unannounced inspections, review police records, and interview personnel and civilians.

    - **Reporting Mechanisms**: Create clear and accessible reporting mechanisms for victims and witnesses of human rights violations by police. This could include hotlines, online platforms, and dedicated offices in police stations.

    - **Data Collection and Analysis**: Collect and analyze data on reported human rights violations to identify patterns, hotspots, and recurring issues. Use this data to inform policy changes and training needs.

    - **Public Transparency**: Regularly publish reports on the findings of the monitoring body, including statistics on human rights violations, disciplinary actions taken, and improvements made.

- **Expected Outcomes**:
    - Increased transparency and accountability in law enforcement operations.
    - Enhanced public trust in the police force.
    - Reduction in human rights violations by police and security forces.

Engage Civil Society and Communities

- Actions:
    - **Partnerships with NGOs**: Partner with human rights NGOs and community organizations to provide additional training and support to police forces. These partnerships can help bring in diverse perspectives and expertise.
    - **Community Policing Initiatives**: Implement community policing initiatives that involve local communities in policing efforts. Encourage regular meetings between police officers and community members to discuss safety concerns and build mutual trust.
    - **Public Awareness Campaigns**: Conduct public awareness campaigns to inform citizens about their rights and how to report police misconduct.

Use media, social networks, and community events to reach a broad audience.

- **Feedback Forums**: Organize regular forums where community members can provide feedback directly to police officers and discuss ways to improve community-police relations.

- **Expected Outcomes**:

  - Stronger collaboration and trust between the police and local communities.

  - Increased public awareness of human rights and police accountability mechanisms.

  - More effective and community-oriented policing.

Conclusion: By implementing these detailed actions, the Ministry of Internal Affairs can significantly enhance the respect for human rights among police and security forces in Sierra Leone. Developing comprehensive training programs, conducting regular assessments, establishing monitoring and reporting systems, and engaging civil society and communities are crucial steps. These efforts will foster increased awareness, improved behavior, and greater accountability within law enforcement. Ultimately, this will build a transparent, accountable, and community-focused law enforcement system, enhancing public trust and cooperation and contributing to a safer and more just society.

3. National Electoral Commission

- **Focus**: Ensure free, fair, and transparent electoral processes.

Enhance Voter Education

- Actions:
    - **Voter Education Campaigns**:
        - **Develop Comprehensive Campaigns**: Create a detailed plan for voter education that includes the rights and responsibilities of voters, the importance of voting, and how to participate in the electoral process. Use a variety of media, including radio, television, social media, and community events, to reach a broad audience.
        - **Educational Materials**: Produce and distribute pamphlets, brochures, and posters in multiple languages to ensure accessibility for all citizens. These materials should cover key topics such as voter registration procedures, the voting process, and the significance of each election type (local, parliamentary, presidential).

- **Community Outreach Programs**: Organize community meetings, workshops, and seminars in collaboration with local leaders and civil society organizations to educate citizens on the electoral process and encourage active participation.

- **School Programs**: Implement civic education programs in schools to teach young people about the democratic process and the importance of voting from an early age.

- **Expected Outcomes**:

  - Increased public awareness and understanding of the electoral process.

  - Higher voter turnout and informed participation in elections.

  - Reduced instances of voter fraud and manipulation due to better-informed voters.

Implement Technology Solutions

- Actions:
    - Electronic Voter Registration:
        - **Biometric Registration Systems**: Implement biometric voter registration systems to accurately capture and verify voter identities, reducing the risk of duplicate registrations and impersonation. Ensure these systems are secure and user-friendly.
        - **Centralized Voter Database**: Develop a centralized voter database that can be accessed and updated in real time, ensuring that voter information is accurate and current. Provide mechanisms for voters to check and update their registration details online.
    - Electronic Voting Machines (EVMs):
        - **Pilot Projects**: Conduct pilot projects in selected regions to test the effectiveness and reliability of EVMs. Use the results to refine the technology and address any issues before nationwide implementation.

- **Training and Public Demonstrations**: Train election officials on the use of EVMs and conduct public demonstrations to familiarize voters with the technology, ensuring transparency and building trust in the system.

- **Real-Time Results Reporting**:

  - **Results Transmission System**: Develop a secure, real-time results transmission system that allows election results to be transmitted electronically from polling stations to central tallying centers. Ensure the system is transparent and accessible to observers and the public.

  - **Online Results Portal**: Create an online portal where election results are published in real-time, allowing the public and media to monitor the progress of vote counting and final results.

- **Expected Outcomes**:

  - Enhanced accuracy and reliability of voter registration and election results.

- Increased public confidence in the integrity of the electoral process.

- Reduced opportunities for electoral fraud and manipulation.

Strengthen Legal Frameworks

- Actions:

  - **Review and Update Electoral Laws**:

    - **Comprehensive Legal Review**: Collaborate with legal experts, civil society organizations, and international bodies to review and update existing electoral laws. Address any gaps or weaknesses that could undermine the fairness and integrity of the electoral process.

    - **Stakeholder Consultations**: Conduct extensive consultations with stakeholders, including political parties, community leaders, and the general public, to ensure that legal reforms are inclusive and reflect the needs and concerns of all segments of society.

- **Stringent Penalties for Electoral Malpractice**:

    - **Enforcement Mechanisms**: Implement strict enforcement mechanisms to ensure that violations of electoral laws are promptly investigated and prosecuted. This includes setting up special courts or tribunals to handle electoral disputes and offenses.

    - **Public Awareness Campaigns**: Conduct campaigns to inform the public about the legal consequences of electoral malpractice and encourage citizens to report any instances of fraud or corruption.

- **Expected Outcomes**:

    - A stronger legal framework that upholds the principles of free and fair elections.

    - Increased deterrence of electoral fraud and malpractice.

    - Enhanced public trust in the electoral system and its fairness.

Engage Independent Observers

- Actions:
  - Invitation to Observers:
    - **National and International Observers**: Extend formal invitations to national and international election observer organizations to monitor the electoral process. Ensure that these observers have unrestricted access to polling stations, counting centers, and other relevant locations.
    - **Training for Observers**: Provide comprehensive training for election observers on the legal framework, electoral procedures, and best practices for monitoring elections. Ensure that observers are well-prepared to detect and report any irregularities.
  - Transparency and Reporting:
    - **Observer Reports**: Encourage observers to publish detailed reports on their findings, including any incidents of irregularities, recommendations for improvements, and overall assessments of the electoral process.

These reports should be made publicly available and used to inform future electoral reforms.

- **Collaboration with Media**: Work closely with media organizations to ensure that observer reports and findings are widely disseminated and discussed in the public domain. This transparency will help build public confidence in the electoral process.

- **Expected Outcomes**:
  - Increased transparency and accountability in the electoral process.
  - Greater public and international confidence in the integrity of elections.
  - Valuable insights and recommendations for continuous improvement of the electoral system.

Conclusion: By implementing these detailed actions, the National Electoral Commission can significantly enhance the transparency, fairness, and integrity of the electoral process in Sierra Leone. Through comprehensive voter education, advanced technology solutions strengthened legal frameworks, and the engagement of independent observers, the Commission will ensure more accurate, reliable, and trusted elections.

These efforts will lead to increased public awareness, higher voter turnout, reduced electoral fraud, and greater public confidence in the democratic process. Ultimately, this will foster a more democratic society where every citizen can participate in free and fair elections, strengthening trust in Sierra Leone's political system.

4. Anti-Corruption Commission

- **Focus**: Strengthen anti-corruption laws and enforcement mechanisms.

- **Implementation**:

Enforce Strict Reporting and Accountability Measures for Public Officials

- Actions:
    - **Comprehensive Legal Framework**: Develop and implement robust anti-corruption laws that clearly define corrupt practices and prescribe stringent penalties. Ensure these laws are in line with international standards and best practices.
    - **Mandatory Asset Declaration**: Require all public officials to regularly declare their assets, liabilities, and income. These declarations should be verified by an independent body and made accessible to the public.

- **Regular Audits and Inspections**: Conduct regular financial and performance audits of government agencies and officials. Utilize both internal auditors and external auditors independent auditors to ensure objectivity and thoroughness.

- **Whistleblower Protection**: Establish and enforce laws that protect whistleblowers from retaliation. Create safe and anonymous channels for reporting corruption.

- **Public Access to Information**: Mandate that all government transactions and decisions, especially those involving public funds, be made publicly accessible. This includes contracts, procurement processes, and budget allocations.

- **Expected Outcomes**:
  - Enhanced transparency and accountability of public officials.
  - Reduction in instances of corruption due to increased oversight and stricter penalties.
  - Greater public trust in government institutions.

Launch Public Awareness Campaigns on Transparency and Anti-Corruption Efforts

- Actions:
    - **Education and Outreach Programs**: Develop comprehensive public education programs about the dangers of corruption and the importance of transparency. Use various media, including TV, radio, social media, and community outreach, to reach a broad audience.
    - **Collaborate with Civil Society**: Partner with NGOs, community groups, and other civil society organizations to spread awareness about anti-corruption measures and educate the public on how to report corrupt practices.
    - **Youth Engagement**: Implement educational programs in schools and universities to teach young people about integrity, ethics, and the impact of corruption. Encourage youth participation in anti-corruption activities.
    - **National Anti-Corruption Week**: Organize an annual National Anti-Corruption Week with events, seminars, and workshops focused on combating corruption and promoting transparency.

- **Expected Outcomes:**
  - Increased public awareness and understanding of corruption and its impacts.
  - Greater public involvement in reporting and combating corruption.
  - A cultural shift towards integrity and transparency, particularly among young people.

Establish Effective Investigation and Prosecution Mechanisms

- **Actions:**
  - **Specialized Anti-Corruption Units**: Set up dedicated anti-corruption units within law enforcement agencies to focus exclusively on investigating and prosecuting corruption cases.
  - **Training for Investigators and Prosecutors**: Provide specialized training for investigators and prosecutors on the latest techniques and best practices in handling corruption cases. Include training on financial forensics, digital evidence collection, and international cooperation.

- **Collaboration with International Bodies**: Partner with international anti-corruption bodies and law enforcement agencies to share information, resources, and best practices. Engage in joint investigations where necessary.

- **Streamlined Legal Processes**: Reform legal procedures to ensure swift and efficient prosecution of corruption cases. This includes setting up special anti-corruption courts and fast-tracking corruption-related trials.

- Expected Outcomes:
  - More effective investigation and prosecution of corruption cases.
  - Increased deterrence against corrupt practices due to the likelihood of detection and prosecution.
  - Improved international cooperation and alignment with global anti-corruption standards.

Implement Robust Monitoring and Evaluation Systems

- Actions:
  - **Anti-Corruption Scorecards**: Develop and implement anti-corruption scorecards to regularly assess the performance of government agencies and officials in preventing and combating corruption.

These scorecards should be publicly available.

- **Regular Reporting**: Mandate regular reporting by government agencies on their anti-corruption efforts and the outcomes of these efforts. These reports should be audited and published for public scrutiny.

- **Citizen Feedback Mechanisms**: Establish platforms for citizens to provide feedback on government services and report any instances of corruption. This could include online portals, hotlines, and community feedback sessions.

- **Impact Assessment Studies**: Conduct periodic impact assessment studies to evaluate the effectiveness of anti-corruption measures and identify areas for improvement.

- **Expected Outcomes**:

    - Continuous monitoring and improvement of anti-corruption efforts.

    - Increased transparency and accountability in government operations.

    - Better-informed and engaged citizenry in the fight against corruption.

Conclusion: By implementing these detailed actions, the Anti-Corruption Commission can significantly strengthen the legal and institutional framework for combating corruption in Sierra Leone. Enforcing strict reporting and accountability measures, launching public awareness campaigns, establishing effective investigation and prosecution mechanisms, and implementing robust monitoring systems will enhance transparency, accountability, and integrity in public office. These efforts will reduce corruption, increase public trust in government institutions, and foster a culture of zero tolerance for corruption. Ultimately, this will promote sustainable development and good governance, ensuring a fairer and more just society for all Sierra Leoneans.

5. Ministry of Finance

- **Focus**: Implement transparent budgetary processes.
- **Implementation**:

Develop Clear and Accessible Financial Reports for Public Scrutiny

- Actions:
    - **Standardized Reporting Formats**: Create standardized formats for financial reports to ensure consistency and clarity. These reports should detail revenue collections, expenditures, budget allocations, and financial performance indicators.

- **Regular Publication of Financial Data**: Mandate the regular publication of detailed financial reports on the Ministry of Finance's website and through other public channels. Ensure these reports are updated quarterly and annually.

- **Open Data Portals**: Develop and maintain an open data portal where citizens can access comprehensive financial data. This portal should be user-friendly and provide tools for data analysis and visualization.

- **Community Outreach and Education**: Conduct community outreach programs and workshops to educate the public on how to interpret financial reports and use the open data portal. This will empower citizens to engage more effectively in budget monitoring and advocacy.

- **Collaboration with Civil Society**: Partner with civil society organizations and academic institutions to analyze financial data, produce independent budget analyses, and disseminate findings to the public.

- **Expected Outcomes**:
    - Increased transparency and accountability in the management of public finances.
    - Enhanced public understanding of government spending and revenue collection.
    - Greater public trust in the Ministry of Finance and the government's financial management practices.

Optimize Resource Allocation for Maximum Effectiveness and Efficiency

- **Actions**:
    - **Performance-Based Budgeting**: Implement a performance-based budgeting system that links budget allocations to specific outcomes and performance indicators. This approach ensures that funds are allocated based on the effectiveness and efficiency of programs and projects.
    - **Expenditure Reviews**: Conduct regular expenditure reviews to assess the efficiency and effectiveness of government spending. Use these reviews to identify areas of waste, overlap, and opportunities for cost savings.

- **Priority Setting and Planning**: Establish clear criteria for setting budget priorities based on national development goals, socio-economic needs, and public feedback. Engage in strategic planning to align budget allocations with long-term development objectives.

- **Integrated Financial Management Information Systems (IFMIS)**: Implement and maintain an integrated financial management information system to improve financial planning, budgeting, accounting, and reporting. Ensure the system is secure, reliable, and accessible to relevant stakeholders.

- **Public Investment Management**: Strengthen the framework for public investment management to ensure that projects are selected, prioritized, and implemented based on rigorous cost-benefit analysis and socio-economic impact assessments.

- **Expected Outcomes**:
  - More efficient use of public resources and reduced wasteful spending.
  - Improved alignment of budget allocations with national priorities and development goals.

- Enhanced accountability and performance monitoring of government programs and projects.

Enhance Transparency in Procurement Processes

- Actions:
    - **Public Procurement Reforms**: Develop and implement comprehensive public procurement reforms to ensure transparency, competitiveness, and integrity in the procurement process. This includes establishing clear guidelines and procedures for procurement planning, bidding, evaluation, and contract management.

    - **E-Procurement System**: Implement an electronic procurement (e-procurement) system to streamline procurement processes, reduce opportunities for corruption, and increase transparency. Ensure the system is user-friendly and accessible to all potential bidders.

    - **Capacity Building for Procurement Officials**: Provide regular training for procurement officials on best practices in public procurement, ethical standards, and the use of e-procurement systems. Ensure that officials are equipped with the skills and knowledge to conduct transparent and efficient procurement processes.

- **Independent Procurement Oversight**: Establish an independent body to oversee and audit public procurement processes. This body should have the authority to investigate complaints, conduct audits, and enforce compliance with procurement regulations.

- **Stakeholder Engagement**: Engage with stakeholders, including suppliers, contractors, civil society organizations, and the public, to gather feedback on procurement processes and address any concerns. Promote dialogue and collaboration to improve procurement practices and outcomes.

Expected Outcomes:

- Increased transparency and accountability in public procurement.

- Reduced opportunities for corruption and fraud in the procurement process.

- Improved efficiency and effectiveness of procurement activities, leading to better value for money in public spending.

Implement Robust Monitoring and Evaluation Systems

- Actions:
    - **Budget Monitoring Committees**: Establish budget monitoring committees at national and local levels to oversee the implementation of budget allocations and expenditures. These committees should include representatives from government, civil society, and the private sector.
    - **Performance Audits**: Conduct regular performance audits to assess the effectiveness and efficiency of government programs and projects. Use audit findings to inform budget adjustments and policy decisions.
    - **Citizen Participation in Budget Oversight**: Create mechanisms for citizen participation in budget oversight, such as public hearings, participatory budgeting initiatives, and feedback forums. Encourage citizens to actively engage in monitoring government spending and holding officials accountable.
    - **Annual Budget Reports**: Publish comprehensive annual budget reports that detail budget performance, key achievements, challenges, and areas for improvement.

Ensure these reports are accessible to the public and widely disseminated.

- **Feedback and Complaints Mechanisms**: Establish clear mechanisms for receiving and addressing feedback and complaints related to budget implementation and financial management. Ensure that these mechanisms are accessible and responsive to citizens' concerns.

- **Expected Outcomes**:

    - Enhanced accountability and transparency in budget implementation and financial management.

    - Improved efficiency and effectiveness of government programs and projects.

    - Greater public trust and confidence in the Ministry of Finance and the government's fiscal policies.

Conclusion: By implementing these detailed actions, the Ministry of Finance can significantly enhance the transparency, accountability, and efficiency of its financial management processes. Developing clear and accessible financial reports, optimizing resource allocation, enhancing transparency in procurement, and establishing robust monitoring and evaluation systems are crucial steps. These efforts will lead to more effective use of public resources, reduced wasteful spending, and improved public trust in government institutions.

Ultimately, this will contribute to sustainable socio-economic development in Sierra Leone, ensuring that government finances are managed responsibly and transparently, benefiting all citizens.

6. Ministry of Education

- **Focus**: Integrate civic education into school curricula.
- **Implementation**:

Focus: Integrate Civic Education into School Curricula

Implementation:

Develop Comprehensive Civic Education Curricula

- Actions:
    - **Curriculum Design**: Collaborate with educational experts, civil society organizations, and international bodies to design a comprehensive civic education curriculum. This curriculum should cover democratic principles, human rights, the responsibilities of citizens, and the importance of civic participation.
    - **Incorporate Local Context**: Ensure that the curriculum includes local historical and cultural contexts to make it relevant and engaging for students. Highlight significant events and figures in Sierra Leone's history that have shaped its democratic journey.

- **Interactive Teaching Methods**: Develop interactive and participatory teaching methods, such as debates, role-plays, simulations, and community projects, to make civic education more engaging and impactful.

- **Resource Development**: Produce textbooks, teaching guides, multimedia resources, and other educational materials that support the civic education curriculum. Ensure these resources are available in multiple languages and accessible to all students.

- **Expected Outcomes**:
  - Increased awareness and understanding of democratic principles and civic responsibilities among students.
  - Enhanced critical thinking, participation, and engagement in democratic processes from a young age.
  - A generation of informed and active citizens committed to promoting and upholding democratic values.

Train Teachers and Educators

- **Actions:**

    - **Professional Development Programs:** Establish professional development programs for teachers and educators focused on civic education. These programs should include workshops, seminars, and certification courses to equip teachers with the knowledge and skills to effectively teach civic education.

    - **Continuous Support and Resources:** Provide continuous support to teachers through resource centers, online platforms, and regular updates on best practices in civic education. Create networks and communities of practice where educators can share experiences and resources.

    - **Incentives for Participation:** Develop incentive schemes to encourage teachers to participate in professional development programs and to innovate in their teaching methods. Recognize and reward outstanding teachers who excel in teaching civic education.

- **Expected Outcomes:**

    - A well-trained and motivated cadre of teachers capable of delivering high-quality civic education.

- Enhanced teaching quality and effectiveness in civic education across schools.

- Increased student engagement and performance in civic education.

Integrate Civic Education into Extracurricular Activities

- Actions:

    - **Student Clubs and Organizations**: Promote the formation of student clubs and organizations focused on civic engagement, human rights, and community service. Support these clubs with resources, mentorship, and opportunities for involvement in local governance and community projects.

    - **Civic Education Events**: Organize events such as mock elections, debates, public speaking competitions, and community service projects to provide students with practical experiences in civic participation. Involve local leaders and community members to enhance the impact and relevance of these events.

    - **School-Community Partnerships**: Foster partnerships between schools and local communities to implement joint civic education initiatives.

> Encourage students to participate in community meetings, local government activities, and civic awareness campaigns.

- **Expected Outcomes**:
  - Increased practical understanding of civic concepts among students through hands-on experiences.
  - Strengthened school-community relationships and greater community involvement in education.
  - Enhanced leadership and civic participation skills among students.

Monitor and Evaluate Civic Education Programs

- **Actions**:
  - **Assessment Frameworks**: Develop and implement assessment frameworks to evaluate the effectiveness of civic education programs. Use a mix of formative and summative assessments, including student feedback, teacher evaluations, and performance metrics.
  - **Regular Reviews and Adjustments**: Conduct regular reviews of the civic education curriculum and teaching methods based on assessment data.

Adjust and update the curriculum and teaching strategies to address identified gaps and improve outcomes.

- **Impact Studies**: Commission independent studies to assess the long-term impact of civic education on students' attitudes, knowledge, and behaviors regarding democratic participation and civic responsibility.

- **Expected Outcomes**:
    - Continuous improvement in the quality and effectiveness of civic education programs.
    - Data-driven decision-making and curriculum adjustments based on assessment results.
    - Enhanced ability to measure and demonstrate the impact of civic education on student outcomes and democratic participation.

Conclusion: By implementing these detailed actions, the Ministry of Education can significantly enhance the integration of civic education into school curricula. Developing a comprehensive curriculum, training teachers, incorporating extracurricular activities, and monitoring and evaluating programs are essential steps.

These efforts will lead to increased awareness and understanding of democratic principles among students, enhanced critical thinking and civic engagement, and the development of a generation committed to upholding democratic values. Ultimately, this will contribute to the formation of informed, engaged, and responsible citizens who actively participate in the governance and democratic processes of Sierra Leone.

7. Ministry of Information and Communications

- **Focus**: Ensure freedom of the press and promote transparent government communication.

- **Implementation**:

Ensure Freedom of the Press

- Actions:
  - **Legal Frameworks for Press Freedom**: Review and strengthen existing laws to protect press freedom, ensuring they align with international standards such as Article 19 of the Universal Declaration of Human Rights and the African Charter on Human and Peoples' Rights.
  - **Abolish Restrictive Laws**: Identify and repeal laws that hinder press freedom, such as those imposing excessive penalties for defamation or restricting access to information.

- **Independent Regulatory Bodies**: Establish or strengthen independent media regulatory bodies to oversee media practices, ensuring they operate without political interference and promote ethical journalism standards.

- **Support for Journalists**: Provide legal and financial support for journalists who face harassment or legal challenges due to their work. Establish a legal defense fund and offer training on their rights and how to protect themselves legally and physically.

- **Press Freedom Index**: Develop a national press freedom index to monitor and report on the state of press freedom in Sierra Leone annually. Use the findings to inform policy and advocacy efforts.

- **Expected Outcomes**:
  - Enhanced protection for journalists and media organizations.
  - Increased independence and diversity in the media landscape.
  - Improved public trust in media as a source of reliable information.

Promote Transparent Government Communication

- Actions:
    - **Government Transparency Policies**: Develop and implement policies that mandate transparency in all government communications. This includes the proactive disclosure of information related to government activities, budgets, decisions, and public services.
    - **Open Data Initiatives**: Launch open data initiatives where government data is made available to the public in easily accessible formats. This includes data on government spending, public procurement, and development projects.
    - **Regular Press Briefings**: Conduct regular press briefings and provide timely updates on government activities, policies, and decisions. Ensure that these briefings are accessible to all media outlets and the public.
    - **Government Websites and Portals**: Develop and maintain comprehensive government websites and portals that provide up-to-date information and resources. Ensure these platforms are user-friendly and regularly updated.

- **Social Media Engagement**: Utilize social media platforms to engage with citizens, disseminate information, and gather feedback. Use these platforms to promote transparency and facilitate two-way communication between the government and the public.

- **Expected Outcomes**:
  - Increased transparency and accountability of government actions.
  - Enhanced public access to information.
  - Greater public trust and engagement with government initiatives.

Build Capacity for Effective Communication

- **Actions**:
  - **Training Programs for Government Officials**: Implement training programs for government officials on effective communication strategies, transparency, and the use of digital tools. Focus on building skills in public speaking, media relations, and crisis communication.
  - **Public Information Campaigns**: Develop and execute public information campaigns to educate citizens about government programs, services, and their rights.

Use multiple channels, including radio, TV, social media, and community outreach.

- **Feedback Mechanisms**: Establish mechanisms for citizens to provide feedback on government communications and services. This can include online surveys, suggestion boxes, and public forums. Ensure that feedback is reviewed and used to improve communication strategies.

- **Partnerships with Media Organizations**: Collaborate with media organizations to ensure accurate and balanced reporting on government activities. Provide journalists with access to information and resources to facilitate informed reporting.

- **Expected Outcomes**:
  - Improved communication skills among government officials.
  - More effective and transparent dissemination of information to the public.
  - Enhanced public engagement and satisfaction with government services.

Monitor and Evaluate Communication Efforts

- Actions:

    - **Performance Metrics**: Develop performance metrics to evaluate the effectiveness of government communication efforts. Metrics should include public access to information, media coverage, citizen engagement, and feedback.

    - **Regular Audits and Reports**: Conduct regular audits of government communication channels and strategies. Publish reports on the findings and use them to inform improvements.

    - **Citizen Surveys**: Conduct periodic surveys to gauge public perception of government transparency and communication. Use survey results to identify strengths and areas for improvement.

    - **Continuous Improvement**: Establish a process for continuous improvement of communication strategies based on audit findings, performance metrics, and citizen feedback.

- **Expected Outcomes**:
  - Continuous enhancement of government communication strategies.
  - Increased public satisfaction and trust in government communications.
  - More responsive and adaptive communication practices.

Conclusion: By implementing these detailed actions, the Ministry of Information and Communications can significantly enhance press freedom, government transparency, and effective communication in Sierra Leone. Strengthening legal frameworks, abolishing restrictive laws, and supporting journalists will protect press freedom. Promoting transparent government communication through policies, open data initiatives, and regular updates will increase accountability. Building capacity for effective communication and continuously monitoring efforts will improve public access to information and trust in government. These initiatives will foster a more informed, engaged, and empowered citizenry, enhancing collaboration and trust between the government and the public, and promoting democratic governance.

8. Ministry of Local Government and Rural Development

- **Focus**: Strengthen local governance structures.

- **Implementation**:

Facilitate Inclusive Dialogues Between Local Authorities and Communities

- Actions:
  - **Community Engagement Forums**: Organize regular community engagement forums where local authorities and community members can discuss local issues, development plans, and governance matters. Ensure these forums are inclusive, giving voice to women, youth, and marginalized groups.

  - **Public Consultations**: Conduct public consultations before implementing major policies or projects. This involves holding town hall meetings, focus groups, and public hearings to gather input and feedback from the community.

  - **Participatory Budgeting**: Introduce participatory budgeting processes where community members can directly influence how local funds are allocated. This enhances transparency and ensures that resources are directed towards areas of greatest need.

- **Local Development Committees**: Establish local development committees comprising elected community representatives and local government officials. These committees should meet regularly to discuss and plan local development initiatives.

- **Feedback Mechanisms**: Implement mechanisms for collecting ongoing feedback from the community, such as suggestion boxes, online platforms, and community liaisons. Ensure feedback is reviewed and acted upon promptly.

- **Expected Outcomes**:
    - Enhanced community participation in local governance.
    - Greater alignment of local government actions with community needs and priorities.
    - Increased trust and cooperation between local authorities and communities.

Ensure Local Governance Structures are Responsive to Community Needs

- Actions:

    - **Needs Assessment Surveys**: Conduct regular needs assessment surveys to identify the most pressing issues and priorities within communities. Use the data collected to inform policy decisions and resource allocation.

    - **Service Delivery Standards**: Develop and enforce clear standards for local government service delivery. These standards should outline expected levels of service, response times, and quality benchmarks.

    - **Capacity Building for Local Officials**: Provide training and capacity-building programs for local government officials on responsive governance, community engagement, and service delivery. Focus on skills such as project management, conflict resolution, and participatory planning.

    - **Decentralized Decision-Making**: Promote decentralized decision-making by empowering local authorities to make decisions and allocate resources independently. Ensure that decision-making processes are transparent and inclusive.

- **Citizen Report Cards**: Implement citizen report cards to gather feedback on local government performance and service delivery. Use the feedback to identify areas for improvement and hold officials accountable.

- **Expected Outcomes**:

    - Improved responsiveness of local governance structures to community needs.

    - Higher quality and more timely delivery of public services.

    - Empowered local authorities capable of addressing local issues effectively.

Strengthen Local Governance Accountability and Transparency

- **Actions**:

    - **Transparent Budget Processes**: Implement transparent budgeting processes at the local level. Ensure that budget proposals, allocations, and expenditures are publicly accessible and clearly communicated to the community.

    - **Regular Audits and Public Reporting**: Conduct regular financial and performance audits of local government activities.

Publish audit reports and other relevant information in an accessible format to promote accountability.

- **Accountability Mechanisms**: Establish clear accountability mechanisms for local government officials, including performance reviews, public disclosure of assets, and disciplinary actions for misconduct. Ensure these mechanisms are robust and enforced consistently.

- **Community Monitoring Groups**: Create community monitoring groups to oversee local government projects and expenditures. Provide these groups with the training and resources needed to conduct effective oversight.

- **Whistleblower Protections**: Implement strong protections for whistleblowers who report corruption, mismanagement, or other misconduct within local governance structures. Ensure that whistleblowers are protected from retaliation and their reports are investigated thoroughly.

- **Expected Outcomes**:

  - Increased transparency and accountability in local governance.

  - Reduced instances of corruption and mismanagement.

- Greater public trust and confidence in local government institutions.

Promote Sustainable Development and Inclusive Growth in Rural Areas

- Actions:
    - **Integrated Rural Development Plans**: Develop and implement integrated rural development plans that address key areas such as agriculture, infrastructure, education, health, and economic development. Ensure these plans are participatory and reflect the needs of rural communities.
    - **Public-Private Partnerships (PPPs)**: Foster public-private partnerships to leverage additional resources and expertise for rural development projects. Encourage private sector investment in infrastructure, agriculture, and other critical sectors.
    - **Support for Local Enterprises**: Provide support for local enterprises through training, access to finance, and market linkages. Promote entrepreneurship and small business development as key drivers of rural economic growth.

- **Sustainable Practices**: Promote sustainable practices in agriculture, resource management, and infrastructure development. Encourage the adoption of environmentally friendly technologies and practices to ensure long-term sustainability.

- **Infrastructure Development**: Invest in the development of critical infrastructure in rural areas, such as roads, water supply, sanitation, electricity, and telecommunications. Ensure that infrastructure projects are planned and implemented in consultation with the community.

- **Expected Outcomes**:
    - Enhanced economic opportunities and improved living standards in rural areas.
    - Sustainable development practices that protect the environment and natural resources.
    - Strengthened local economies and reduced rural-urban migration.

Conclusion: By implementing these detailed actions, the Ministry of Local Government and Rural Development can significantly strengthen local governance structures, promote responsive and transparent governance, and drive sustainable development in rural areas. Facilitating inclusive dialogues, ensuring responsiveness to community needs, enhancing accountability, and promoting sustainable development are crucial steps. These efforts will lead to greater community participation, improved service delivery, reduced corruption, and strengthened local economies. Ultimately, this will contribute to more inclusive and equitable growth, improved living standards, and greater public trust in local government institutions, fostering a more just and prosperous Sierra Leone.

9. Parliament

- **Focus**: Ensure legislative processes are transparent and involve public consultation.

- **Implementation**:

Establish Public Forums for Legislative Reviews and Feedback

- Actions:
    - **Regular Public Consultations**: Conduct regular public consultations on proposed legislation. This involves organizing town hall meetings, online forums, and focus groups to gather input from citizens, civil society organizations, and other

stakeholders.

- **Legislative Hearings**: Make legislative hearings open to the public and widely publicized. Provide opportunities for citizens and experts to testify and present their views on proposed laws.

- **Draft Legislation Publication**: Publish draft legislation on the parliamentary website and other accessible platforms well in advance of debates and votes. Encourage public comments and suggestions on these drafts.

- **Community Outreach Programs**: Implement community outreach programs to educate the public on the legislative process and the importance of their participation. Use multiple channels to reach diverse audiences, including social media, radio, and community events to reach diverse audinces.

- **Feedback Mechanisms**: Develop mechanisms for continuous feedback on the legislative process, such as suggestion boxes, online surveys, and dedicated hotlines. Ensure that all feedback is reviewed and considered in the legislative process.

- **Expected Outcomes**:
    - Increased public engagement and input in the legislative process.
    - Greater transparency and accountability in legislative decision-making.
    - Legislation that better reflects the needs and concerns of the public.

Maintain Transparent Documentation of Legislative Proceedings

- **Actions**:
    - **Live Broadcasts**: Broadcast parliamentary sessions live on television, radio, and online platforms to allow citizens to follow legislative proceedings in real time.
    - **Archived Records**: Maintain a comprehensive online archive of parliamentary proceedings, including video recordings, transcripts, and minutes of meetings. Ensure these records are easily accessible to the public.
    - **Detailed Legislative Reports**: Publish detailed reports on legislative activities, including the progress of bills, voting records, committee reports, and debates. Provide summaries and analyses to help the public understand complex legislative issues.

- **Transparency Portals**: Develop and maintain transparency portals where all legislative documents, including bills, amendments, reports, and decisions, are published and regularly updated.

- **Open Data Initiatives**: Implement open data initiatives to provide public access to parliamentary data in machine-readable formats. Encourage the use of this data by researchers, journalists, and civil society organizations for analysis and advocacy.

- **Expected Outcomes**:

  - Enhanced public access to information on legislative activities.

  - Improved transparency and accountability in the legislative process.

  - Increased public trust in parliamentary institutions.

Facilitate Citizen Participation in Legislative Processes

- Actions:

  - **Participatory Platforms**: Develop and promote platforms for citizen participation, such as e-petitions, online consultations, and participatory budgeting initiatives. Ensure these platforms are user-friendly and accessible to all citizens.

- **Citizen Advisory Committees**: Establish citizen advisory committees to provide input on key legislative issues. These committees should include representatives from diverse sectors of society, including marginalized and underrepresented groups.

- **Educational Campaigns**: Conduct educational campaigns to raise awareness about the importance of civic participation and how citizens can get involved in the legislative process. Use schools, community centers, and media outlets to disseminate information.

- **Youth Engagement Programs**: Implement programs to engage young people in the legislative process, such as mock parliaments, internships, and mentorship programs. Encourage youth participation in public consultations and legislative reviews.

- **Accessibility Initiatives**: Ensure that all legislative processes and materials are accessible to people with disabilities. This includes providing sign language interpretation, captioning, and documents in Braille and other accessible formats.

- **Expected Outcomes**:
  - Increased citizen participation and engagement in legislative processes.
  - Legislation that is more inclusive and reflective of diverse perspectives.
  - Empowered citizens who are informed and actively involved in governance.

Strengthen Oversight and Accountability Mechanisms

- **Actions**:
  - **Parliamentary Committees**: Strengthen the role of parliamentary committees in overseeing the executive branch and holding it accountable. Ensure committees have the resources and authority to conduct thorough investigations and hearings.
  - **Performance Audits**: Conduct regular performance audits of government programs and agencies. Use the findings to hold the executive accountable and to inform legislative decisions.
  - **Question Time**: Institutionalize regular question time sessions where parliamentarians can question government officials on their actions and policies. Ensure these sessions are open to the public and widely covered by the media.

- **Corruption Investigations**: Establish mechanisms for investigating and addressing allegations of corruption and misconduct within the legislative branch. This includes setting up independent bodies to oversee and enforce ethical standards.

- **Public Reporting**: Publish detailed reports on the outcomes of oversight activities, including findings from audits, investigations, and hearings. Ensure these reports are accessible to the public and widely disseminated.

- **Expected Outcomes**:
  - Enhanced oversight and accountability of the executive branch.
  - Reduced instances of corruption and misconduct in government.
  - Increased public confidence in the integrity and effectiveness of parliamentary oversight.

Conclusion: By implementing these detailed actions, Parliament can significantly enhance the transparency, accountability, and inclusiveness of its legislative processes. Establishing public forums for legislative reviews, maintaining transparent documentation, facilitating citizen participation, and strengthening oversight mechanisms are crucial steps.

These efforts will lead to increased public engagement, greater accountability in legislative decision-making, and improved public trust in parliamentary institutions. Ultimately, this will contribute to more effective governance, a stronger democratic system, and legislation that better reflects the needs and concerns of the people in Sierra Leone, fostering a more just and participatory society.

10. Judiciary

- **Focus**: Promote judicial reforms for impartiality and integrity.
- **Implementation**:

Strengthen Judicial Independence

- **Actions**:
    - **Constitutional and Legal Reforms**: Amend the constitution and relevant laws to reinforce the independence of the judiciary. Ensure that judicial appointments, promotions, and removals are based on merit and free from political influence.
    - **Independent Judicial Service Commission**: Establish or strengthen an independent Judicial Service Commission (JSC) responsible for overseeing the appointment, promotion, and discipline of judges.

Ensure that the JSC operates transparently and includes members from diverse sectors, including legal professionals, academia, and civil society.

- **Secure Tenure for Judges**: Guarantee secure tenure for judges to protect them from arbitrary dismissal or political pressure. Ensure that judges can only be removed through a transparent and fair process for proven misconduct or incapacity.

- **Adequate Remuneration and Resources**: Ensure that judges and judicial staff receive adequate remuneration and resources to perform their duties effectively. This includes competitive salaries, benefits, and funding for court infrastructure and administrative support.

- **Expected Outcomes**:

    - Enhanced judicial independence and impartiality.

    - Increased public trust in the judiciary as a fair and unbiased institution.

    - Reduced political interference in judicial matters.

Enhance Transparency and Accountability in the Judiciary

- Actions:

  - **Public Access to Judicial Decisions**: Mandate that all judicial decisions are published online and made accessible to the public. Develop a user-friendly database where citizens can search and view court rulings, judgments, and legal precedents.

  - **Performance Monitoring and Reporting**: Implement systems for monitoring and reporting the performance of judges and courts. This includes tracking case backlog, timeliness of judgments, and adherence to procedural standards. Publish regular performance reports to maintain transparency.

  - **Complaint and Disciplinary Mechanisms**: Establish clear and accessible mechanisms for lodging complaints against judges and judicial staff. Ensure that complaints are investigated promptly and fairly by an independent body, and take appropriate disciplinary actions when necessary.

  - **Asset Declaration**: Require judges to regularly declare their assets and make these declarations publicly available.

This promotes transparency and helps prevent corruption within the judiciary.

- **Expected Outcomes**:

    - Increased transparency and accountability within the judiciary.

    - Improved efficiency and performance of judicial processes.

    - Greater public confidence in the integrity and effectiveness of the judiciary.

Improve Access to Justice

- Actions:

    - **Legal Aid Services**: Expand and strengthen legal aid services to ensure that all citizens, especially those from marginalized and disadvantaged communities, have access to legal representation. This includes funding for legal aid clinics and support for pro bono legal work by private practitioners.

    - **Simplified Procedures**: Simplify court procedures to make them more accessible and understandable to the general public. Provide clear guidelines and assistance for individuals representing themselves in court (pro se litigants).

- **Mobile Courts and Outreach Programs**: Establish mobile courts and legal outreach programs to bring justice services closer to remote and underserved communities. Conduct regular legal awareness campaigns to educate citizens about their rights and the judicial process.

- **Alternative Dispute Resolution (ADR)**: Promote and institutionalize alternative dispute resolution mechanisms, such as mediation and arbitration, to provide faster and more cost-effective means of resolving disputes. Ensure that ADR services are accessible and integrated into the formal judicial system.

- **Expected Outcomes**:
  - Enhanced access to justice for all citizens, regardless of socio-economic status.
  - Reduced case backlog and faster resolution of legal disputes.
  - Increased public awareness and understanding of legal rights and judicial processes.

Capacity Building for Judicial Personnel

- Actions:

    - **Continuous Professional Development**: Implement continuous professional development programs for judges, magistrates, and judicial staff. These programs should cover topics such as judicial ethics, case management, human rights, and emerging legal issues.

    - **Training Institutes**: Establish or strengthen judicial training institutes to provide specialized training for new and existing judges. Ensure that these institutes have the resources and expertise to offer high-quality training programs.

    - **International Collaboration and Exchange Programs**: Facilitate collaboration and exchange programs with judicial institutions in other countries. This allows judges and judicial staff to learn from best practices and innovations in different legal systems.

    - **Technology and Innovation**: Train judicial personnel on the use of technology and digital tools to enhance the efficiency and effectiveness of judicial processes. This includes case management systems, electronic filing, and virtual court hearings.

- **Expected Outcomes**:
  - Improved knowledge and skills among judges and judicial staff.
  - Enhanced capacity of the judiciary to handle complex and emerging legal issues.
  - Increased efficiency and effectiveness of judicial processes through the use of technology.

Conclusion: By implementing these detailed actions, the Judiciary can significantly enhance its independence, transparency, accessibility, and capacity. Strengthening judicial independence through constitutional reforms, ensuring transparency with public access to decisions, and improving access to justice with expanded legal aid and mobile courts are essential steps. Additionally, capacity building for judicial personnel through continuous professional development and technological innovation will further enhance efficiency and effectiveness. These efforts will contribute to a more impartial, efficient, and trustworthy judicial system, promoting the rule of law and protecting the rights of all citizens in Sierra Leone, thereby fostering greater public trust and confidence.

11. Human Rights Commission

- **Focus**: Conduct public awareness campaigns on human rights.

- **Implementation**:

Promote and Protect Human Rights Through Educational Initiatives

  - Actions:
    - **Human Rights Curriculum**: Develop and integrate a comprehensive human rights curriculum in schools and universities. This curriculum should cover international human rights standards, national laws, and practical ways to uphold human rights.

    - **Teacher Training**: Provide training for teachers and educators on human rights education. Equip them with the necessary tools and resources to effectively teach human rights topics.

    - **Community Workshops and Seminars**: Organize workshops and seminars in communities to educate citizens about their human rights and how to protect them. These events should be interactive and tailored to the specific needs and concerns of the community.

- **Publications and Resources**: Produce and distribute educational materials such as pamphlets, booklets, and posters that explain human rights in simple terms. Ensure these resources are available in multiple languages and accessible formats.

- **Expected Outcomes**:
  - Increased public awareness and understanding of human rights.
  - Better-equipped educators to teach human rights effectively.
  - Empowered citizens who are aware of their rights and how to protect them.

Monitor and Report on Human Rights Violations

- **Actions**:
  - **Data Collection and Analysis**: Establish robust systems for collecting and analyzing data on human rights violations. This includes setting up hotlines, online reporting tools, and partnerships with local organizations to gather information.
  - **Regular Reporting**: Publish regular reports on the state of human rights in Sierra Leone. These reports should highlight trends, provide detailed accounts of violations, and make recommendations for improvement.

- **Case Investigations**: Investigate reported human rights violations thoroughly and impartially. Work with law enforcement and the judiciary to ensure that perpetrators are held accountable and victims receive justice.

- **Public Hearings**: Conduct public hearings on significant human rights issues. These hearings provide a platform for victims, witnesses, and experts to share their experiences and insights.

- **Expected Outcomes**:
  - Enhanced documentation and understanding of human rights issues.
  - Increased accountability for human rights violations.
  - Improved public confidence in the Human Rights Commission's ability to protect their rights.

Advocate for Policy and Legislative Reforms

- Actions:
  - Legal Review and Recommendations: Conduct regular reviews of existing laws and policies to identify gaps and inconsistencies with international human rights standards. Provide recommendations for reforms to align national legislation with global norms.

- **Lobbying and Advocacy**: Engage in lobbying and advocacy efforts with lawmakers, government officials, and other stakeholders to promote the adoption of human rights-friendly laws and policies.

- **Public Campaigns**: Launch public campaigns to build support for policy and legislative changes. Use media, social networks, and community events to raise awareness and mobilize public opinion.

- **Collaborate with Civil Society**: Work closely with civil society organizations to advocate for human rights reforms. This includes forming coalitions, sharing resources, and coordinating efforts to amplify impact.

- **Expected Outcomes**:
  - Stronger legal and policy frameworks that protect human rights.
  - Increased political will to adopt and implement human rights reforms.
  - Greater public support for human rights initiatives.

Provide Legal and Support Services to Victims of Human Rights Violations

- Actions:
    - **Legal Aid Clinics**: Establish legal aid clinics to provide free legal assistance to victims of human rights violations. These clinics should offer services such as legal advice, representation, and support in navigating the justice system.
    - **Counseling and Support Services**: Provide counseling and support services to victims of human rights violations, including psychological support, social services, and referrals to other support organizations.
    - **Victim Advocacy Programs**: Develop victim advocacy programs that help victims understand their rights, access services, and participate in legal proceedings. Advocate on behalf of victims to ensure they receive fair treatment and justice.
    - **Awareness and Outreach**: Conduct awareness and outreach programs to inform victims of available services and how to access them. Use various communication channels to reach different segments of the population.

- **Expected Outcomes**:
  - Enhanced access to justice and support for victims of human rights violations.
  - Improved recovery and empowerment of victims.
  - Strengthened community trust in the Human Rights Commission's ability to provide assistance.

Strengthen Institutional Capacity

- **Actions**:
  - **Capacity Building Programs**: Implement capacity-building programs for Human Rights Commission staff, focusing on areas such as human rights law, investigative techniques, case management, and advocacy skills.
  - **Resource Allocation**: Ensure the Human Rights Commission has adequate resources, including funding, personnel, and infrastructure, to carry out its mandate effectively.
  - **Partnerships and Collaboration**: Foster partnerships with national and international human rights organizations, government agencies, and other stakeholders to enhance the commission's capacity and effectiveness.

- **Monitoring and Evaluation**: Establish robust monitoring and evaluation systems to assess the commission's performance and impact. Use findings to inform strategic planning and continuous improvement.

- **Expected Outcomes**:
    - Strengthened the capacity of the Human Rights Commission to fulfill its mandate.
    - Enhanced effectiveness and efficiency in protecting and promoting human rights.
    - Increased impact and visibility of human rights initiatives.

Conclusion: By implementing these detailed actions, the Human Rights Commission can significantly enhance its role in promoting and protecting human rights in Sierra Leone. Educational initiatives will raise public awareness and understanding of human rights, while monitoring and reporting will increase accountability. Advocacy efforts will strengthen legal frameworks and providing legal and support services will empower victims. Strengthening institutional capacity will ensure the Commission is well-equipped to fulfill its mandate. These efforts will contribute to a more informed, empowered, and rights-respecting society, fostering a culture of human rights and justice and promoting sustainable development and social harmony in Sierra Leone.

12. Civil Service Commission

- **Focus**: Implement merit-based recruitment and performance management systems.

- **Implementation**:

Develop and Implement Merit-Based Recruitment Processes

- Actions:
  - **Transparent Recruitment Policies**: Establish and enforce clear, transparent, and standardized recruitment policies that emphasize merit and competency. Ensure these policies are publicly available and easily accessible.

  - **Job Descriptions and Specifications**: Develop detailed job descriptions and specifications for all civil service positions. These should outline the required qualifications, skills, and experience needed for each role.

  - **Open and Competitive Examinations**: Introduce open and competitive examinations for civil service entry and promotions. Ensure that these exams are fair, standardized, and aligned with the job requirements.

- **Independent Recruitment Panels**: Form independent recruitment panels to oversee the hiring process. These panels should include representatives from various sectors, including government, civil society, and academia, to ensure impartiality.

- **Online Application Systems**: Implement an online application system to streamline the recruitment process, making it more efficient and accessible. Ensure the system is user-friendly and provides timely updates to applicants.

- **Expected Outcomes**:
  - Increased fairness and transparency in the recruitment process.
  - Higher quality of hires based on merit and competency.
  - Enhanced public trust in the integrity of the civil service recruitment process.

Establish Robust Performance Management Systems

- Actions:
  - **Performance Appraisal Framework**: Develop and implement a comprehensive performance appraisal framework that evaluates civil servants based on clear, objective, and measurable criteria.

This framework should include regular performance reviews, feedback mechanisms, and career development planning.

- **Continuous Training and Development**: Provide continuous training and professional development opportunities for civil servants. Focus on enhancing skills, knowledge, and competencies to improve job performance and career progression.

- **Performance-Based Incentives**: Introduce performance-based incentives and rewards to motivate and recognize high-performing civil servants. This can include promotions, salary increases, bonuses, and public recognition.

- **Accountability Mechanisms**: Establish accountability mechanisms to address underperformance and misconduct. This includes clear disciplinary procedures, performance improvement plans, and, if necessary, termination processes.

- **Performance Monitoring and Reporting**: Implement systems for continuous monitoring and reporting of civil service performance. Use data from performance appraisals to inform management decisions and policy adjustments.

- **Expected Outcomes:**
    - Improved performance and productivity of civil servants.
    - Increased motivation and job satisfaction among civil service employees.
    - Enhanced accountability and professionalism within the civil service.

Promote Ethical Standards and Integrity

- **Actions:**
    - **Code of Conduct**: Develop and enforce a comprehensive code of conduct for all civil servants. This code should outline ethical standards, professional behavior, and the consequences of misconduct.
    - **Ethics Training**: Provide regular ethics training and workshops for civil servants to reinforce the importance of integrity, transparency, and accountability in public service.
    - **Whistleblower Protection**: Establish strong protections for whistleblowers who report unethical behavior, corruption, or misconduct within the civil service. Ensure that whistleblower reports are investigated thoroughly and confidentially.

- **Integrity Assessments**: Conduct regular integrity assessments to identify potential risks and vulnerabilities within the civil service. Use these assessments to develop strategies to mitigate corruption and promote ethical behavior.

- **Expected Outcomes**:

    - Higher ethical standards and integrity within the civil service.

    - Reduced instances of corruption and misconduct.

    - Greater public confidence in the professionalism and accountability of civil servants.

Enhance Inclusivity and Diversity in the Civil Service

- **Actions**:

    - **Equal Opportunity Policies**: Develop and implement equal opportunity policies that promote inclusivity and diversity in civil service recruitment and management. Ensure these policies address gender parity, disability inclusion, and representation of marginalized groups.

    - **Inclusive Recruitment Practices**: Adopt inclusive recruitment practices that actively seek to diversify the civil service workforce.

This includes outreach programs, targeted recruitment drives, and partnerships with diverse organizations.

- **Diversity Training**: Provide diversity and inclusion training for all civil servants to foster an inclusive workplace culture. Focus on raising awareness of unconscious biases, promoting cultural competence, and encouraging respectful interactions.

- **Monitoring and Reporting on Diversity**: Implement systems to monitor and report on diversity within the civil service. Use this data to inform policies and initiatives that promote greater inclusivity and representation.

- **Expected Outcomes**:

    - A more diverse and inclusive civil service workforce.

    - Enhanced representation of marginalized and underrepresented groups.

    - Increased cultural competence and inclusivity within the civil service.

Strengthen Institutional Capacity

- Actions:

    - **Capacity Building Programs**: Implement capacity-building programs to enhance the skills and capabilities of Civil Service Commission member staff. Focus on areas such as human resource management, policy development, data analysis, and leadership.

    - **Resource Allocation**: Ensure that the Civil Service Commission has adequate resources, including funding, personnel, and infrastructure, to effectively carry out its mandate.

    - **Collaboration and Partnerships**: Foster collaboration and partnerships with national and international organizations to share best practices, resources, and expertise. Participate in networks and forums to stay updated on global trends and innovations in civil service management.

    - **Monitoring and Evaluation Systems**: Establish robust monitoring and evaluation systems to assess the commission's performance and impact. Use findings to inform strategic planning and continuous improvement efforts.

- **Expected Outcomes**:
    - Strengthened the capacity and effectiveness of the Civil Service Commission.
    - Improved management and oversight of the civil service.
    - Enhanced ability to implement and sustain reforms.

Conclusion: By implementing these detailed actions, the Civil Service Commission can significantly enhance the transparency, accountability, and effectiveness of civil service management in Sierra Leone. Developing merit-based recruitment processes, establishing robust performance management systems, promoting ethical standards, enhancing inclusivity and diversity, and strengthening institutional capacity are critical steps. These efforts will lead to a more professional, ethical, and inclusive civil service, reducing corruption and improving public trust. Ultimately, this will support the nation's development goals and contribute to a more efficient and effective public administration, fostering a culture of integrity and excellence in public service.

13. Ministry of Gender and Children's Affairs

- **Focus**: Promote gender equality and protect children's rights.

- **Implementation**:

Promote Gender Equality

- Actions:

    - **Legal and Policy Reforms**: Review and amend existing laws and policies to promote gender equality and eliminate discriminatory practices. Ensure compliance with international conventions such as CEDAW (Convention on the Elimination of All Forms of Discrimination Against Women).

    - **Gender Mainstreaming**: Integrate gender perspectives into all government policies, programs, and projects. This includes conducting gender impact assessments and ensuring that gender considerations are part of the planning and implementation processes.

    - **Economic Empowerment Programs**: Develop and implement programs aimed at economically empowering women. This includes providing access to microfinance, entrepreneurship training, vocational skills development, and market linkages.

- **Gender-Based Violence (GBV) Prevention and Response**: Strengthen mechanisms to prevent and respond to gender-based violence. This includes setting up shelters, hotlines, and support services for GBV survivors, as well as training law enforcement and healthcare providers on GBV issues.

- **Public Awareness Campaigns**: Launch public awareness campaigns to challenge gender stereotypes and promote gender equality. Use media, community outreach, and educational programs to change attitudes and behaviors toward women and girls.

- **Expected Outcomes**:
  - Reduced gender disparities in various sectors, including education, employment, and political participation.
  - Increased economic opportunities and financial independence for women.
  - Enhanced protection and support for GBV survivors.
  - Greater public awareness and acceptance of gender equality.

Protect and Promote Children's Rights

- Actions:

  - **Child Protection Policies**: Develop and enforce comprehensive child protection policies that align with international standards such as the UN Convention on the Rights of the Child (CRC). Ensure these policies cover issues such as child labor, trafficking, abuse, and exploitation.

  - **Education Access and Quality**: Promote access to quality education for all children with a focus on marginalized and vulnerable groups. This includes providing scholarships, improving school infrastructure, and training teachers on child-centered pedagogies.

  - **Health and Nutrition Programs**: Implement health and nutrition programs that address the specific needs of children. This includes immunization campaigns, nutritional support, and healthcare services tailored for children.

  - **Child Participation**: Ensure that children's voices are heard in matters that affect them. This includes creating platforms for child participation in decision-making processes at the local and national levels.

- **Support Services for Vulnerable Children**: Provide comprehensive support services for orphans, street children, and children affected by conflict. This includes psychosocial support, family reunification programs, and access to education and healthcare.

- **Expected Outcomes**:
  - Improved legal and policy framework for child protection.
  - Increased access to quality education and healthcare for children.
  - Enhanced protection and support for vulnerable children.
  - Greater involvement of children in decision-making processes.

Capacity Building and Institutional Strengthening

- **Actions**:
  - **Training for Government Officials**: Provide regular training for government officials on gender equality and children's rights. This includes sensitization programs, technical training, and workshops on best practices.

- **Strengthening Data Collection and Analysis**: Improve data collection and analysis on gender and children's issues. This includes developing comprehensive databases and conducting regular surveys and studies to inform policy and program development.

- **Collaboration and Partnerships**: Foster collaboration with national and international organizations, civil society, and the private sector. Establish partnerships to leverage resources, share best practices, and coordinate efforts to promote gender equality and protect children's rights.

- **Monitoring and Evaluation Systems**: Develop robust monitoring and evaluation systems to track the progress and impact of gender equality and child protection initiatives. Use the findings to inform continuous improvement and accountability.

- **Expected Outcomes**:
    - Enhanced capacity of government officials to address gender and children's issues.
    - Improved data and evidence base for policy and program development.

- Strengthened collaboration and coordination among stakeholders.

- Increased accountability and effectiveness of gender equality and child protection initiatives.

Community Engagement and Advocacy

- Actions:

    - **Community-Based Programs**: Implement community-based programs that engage local leaders, community groups, and families in promoting gender equality and protecting children's rights. Focus on culturally sensitive approaches that respect and incorporate local traditions.

    - **Advocacy Campaigns**: Conduct advocacy campaigns to influence public policy and raise awareness about gender and children's issues. Use media, public forums, and social media to reach a broad audience and mobilize support.

    - **Role Models and Champions**: Identify and promote role models and champions for gender equality and children's rights. Encourage influential community members, celebrities, and leaders to advocate for these issues publicly.

- **Capacity Building for Civil Society**: Provide training and support for civil society organizations working on gender and children's issues. Strengthen their capacity to advocate, implement programs, and monitor government actions.

- **Expected Outcomes**:
  - Increased community involvement in promoting gender equality and protecting children's rights.
  - Greater public awareness and support for gender and children's issues.
  - Stronger advocacy efforts and influence on public policy.
  - Enhanced capacity and impact of civil society organizations.

Conclusion: By implementing these detailed actions, the Ministry of Gender and Children's Affairs can significantly enhance the promotion of gender equality and the protection of children's rights in Sierra Leone. Promoting gender equality through legal reforms, economic empowerment, and public awareness campaigns will reduce gender disparities and increase opportunities for women. Protecting children's rights by improving education, healthcare, and support services will ensure their well-being and active participation.

Strengthening institutional capacity and engaging communities will foster collaboration and advocacy, enhancing the overall impact. These efforts will contribute to a more inclusive, equitable, and just society where all individuals, regardless of gender or age, can thrive and reach their full potential.

14. Office of the President

- **Focus**: Lead by example in transparency and accountability.

- **Implementation**:

Establish Transparent Decision-Making Processes

- Actions:
    - **Public Disclosure of Information**: Regularly disclose information on executive decisions, policies, and expenditures. Publish detailed reports on the Office of the President's website, ensuring they are easily accessible to the public.

    - **Transparent Budgeting**: Implement transparent budgeting processes for the Office of the President. Publish the budget, including detailed breakdowns of expenditures, and provide regular updates on budget execution and financial audits.

- **Advisory Councils and Committees**: Create advisory councils and committees that include representatives from civil society, the private sector, and academia. These bodies should provide input on key policy decisions and help ensure that a diverse range of perspectives is considered.

- **Public Consultations**: Conduct public consultations before making major policy decisions. Use town hall meetings, online forums, and surveys to gather input from citizens and stakeholders.

- **Executive Orders and Directives**: Publish all executive orders and directives in a timely manner. Provide clear explanations of the rationale behind each order and its expected impact.

- **Expected Outcomes**:
    - Enhanced transparency in the executive branch.
    - Increased public trust in the Office of the President.
    - Greater accountability in the decision-making process.

Regularly Publish Reports on Government Activities and Expenditures

- Actions:
    - **Annual State of the Nation Reports**: Deliver comprehensive annual State of the Nation reports detailing the government's achievements, challenges, and future plans. Ensure these reports are publicly accessible and widely disseminated.

    - **Quarterly Financial Reports**: Publish quarterly financial reports that provide detailed accounts of government revenue, expenditures, and financial performance. Make these reports available online and in print.

    - **Sectoral Reports**: Issue regular reports on key sectors such as health, education, infrastructure, and security. Highlight progress, challenges, and strategic initiatives in each sector.

    - **Performance Dashboards**: Develop online performance dashboards that track key performance indicators (KPIs) for government programs and projects. Ensure these dashboards are updated regularly and accessible to the public.

- **Independent Audits**: Commission independent audits of government activities and finances. Publish the findings of these audits and take corrective actions as necessary.

- **Expected Outcomes**:
  - Increased public awareness of government activities and financial health.
  - Enhanced accountability through regular public reporting.
  - Improved performance monitoring and management of government programs.

Strengthen Anti-Corruption Measures

- Actions:
  - **Anti-Corruption Policies**: Develop and enforce stringent anti-corruption policies within the executive branch. Ensure these policies cover conflict of interest, gifts and hospitality, and financial disclosures.
  - **Integrity Pacts**: Require all government contracts to include integrity pacts that commit parties to transparent and corruption-free dealings. Monitor compliance and enforce penalties for breaches.

- **Anti-Corruption Task Force**: Establish an anti-corruption task force within the Office of the President to oversee the implementation of anti-corruption measures. Ensure the task force has the authority and resources to investigate and act on corruption allegations.

- **Whistleblower Protections**: Strengthen protections for whistleblowers who report corruption within the executive branch. Ensure that reports are investigated promptly and whistleblowers are protected from retaliation.

- **Public Awareness Campaigns**: Conduct public awareness campaigns to highlight the government's commitment to fighting corruption. Encourage citizens to report corruption and participate in anti-corruption initiatives.

- **Expected Outcomes**:
  - Reduced instances of corruption within the executive branch.
  - Increased public confidence in the government's integrity and commitment to accountability.
  - Enhanced effectiveness of anti-corruption measures.

Foster Inclusivity and Participation in Governance

- Actions:

  - **Inclusive Policy Development**: Ensure that policy development processes are inclusive and participatory. Engage with diverse groups, including women, youth, persons with disabilities, and marginalized communities, to gather their input and perspectives.

  - **National Dialogue Forums**: Organize national dialogue forums to discuss critical issues and gather feedback from citizens. Use these forums to build consensus on major policies and reforms.

  - **Youth and Women's Councils**: Establish youth and women's councils to advise the President on issues affecting these groups. Ensure these councils have a direct line of communication with the Office of the President and can influence policy decisions.

  - **Community Outreach Programs**: Implement community outreach programs to educate citizens about their rights and responsibilities and how they can participate in governance. Use various channels to reach a broad audience, including local media, community meetings, and social media.

- **Civil Society Engagement**: Strengthen partnerships with civil society organizations to enhance their role in monitoring government activities and advocating for citizens' interests. Provide platforms for regular dialogue and collaboration.

- **Expected Outcomes**:
    - Increased inclusivity and representation in governance.
    - Greater citizen engagement and participation in policy-making processes.
    - Strengthened relationships between the government and civil society.

Enhance Capacity and Efficiency of the Executive Branch

- Actions:
    - **Professional Development Programs**: Implement continuous professional development programs for senior government officials. Focus on leadership, strategic planning, public administration, and ethical governance.
    - **Performance Management Systems**: Develop and implement robust performance management systems for the executive branch.

Set clear performance targets, conduct regular evaluations, and provide feedback and incentives for high performance.

- **Technology and Innovation**: Invest in technology and innovation to improve the efficiency of government operations. This includes digital government services, e-governance platforms, and data analytics for decision-making.

- **Organizational Reviews**: Conduct regular organizational reviews to identify and address inefficiencies within the executive branch. Streamline processes, eliminate redundancies, and optimize resource allocation.

- **Collaboration with International Partners**: Foster collaboration with international partners to share best practices, access technical assistance, and leverage resources for capacity-building initiatives.

- **Expected Outcomes**:

  - Enhanced capacity and professionalism within the executive branch.

  - Improved efficiency and effectiveness of government operations.

  - Greater ability to deliver high-quality public services and achieve development goals.

Conclusion: By implementing these detailed actions, the Office of the President can lead by example in promoting transparency, accountability, and inclusivity in governance. Establishing transparent decision-making processes, regularly publishing reports, strengthening anti-corruption measures, fostering inclusivity, and enhancing the capacity of the executive branch are critical steps. These efforts will contribute to a more effective, efficient, and trustworthy executive branch. Increased public trust and citizen engagement will support sustainable development in Sierra Leone and will ensure that government actions are accountable and aligned with the needs and aspirations of the people.

15. Public Procurement Authority

- **Focus**: Ensure fair and transparent procurement processes.

- **Implementation**:

Develop and Enforce Transparent Procurement Guidelines

- Actions:
    - **Procurement Policies and Procedures**: Develop comprehensive procurement policies and procedures that outline clear guidelines for all stages of the procurement process, from planning to contract management. Ensure these guidelines are aligned with international best practices and standards.

- **Standardized Procurement Documents**: Create and mandate the use of standardized procurement documents, including tender notices, bidding documents, contracts, and evaluation criteria. This standardization ensures consistency and transparency across all procurement activities.

- **Procurement Planning and Forecasting**: Implement a robust procurement planning and forecasting system. This system should identify procurement needs in advance, allowing for strategic sourcing and better allocation of resources.

- **Procurement Thresholds and Approval Processes**: Establish clear procurement thresholds and approval processes to ensure that all procurements are subject to appropriate levels of scrutiny and oversight. Define the roles and responsibilities of procurement officials at different levels.

- **Expected Outcomes**:
    - Increased consistency and transparency in procurement processes.
    - Reduced instances of procurement fraud and irregularities.
    - Improved efficiency in procurement planning and execution.

Implement an Electronic Procurement (e-Procurement) System

- Actions:
    - **e-Procurement Platform Development**: Develop and deploy a comprehensive e-procurement platform that facilitates the entire procurement process online. This platform should include modules for supplier registration, tendering, bid submission, evaluation, and contract management.

    - **Training and Capacity Building**: Provide training for procurement officials and suppliers on using the e-procurement platform. Ensure that all stakeholders are familiar with the system's features and functionalities.

    - **User Support and Helpdesk**: Establish a user support and helpdesk service to assist procurement officials and suppliers with technical issues related to the e-procurement system. Provide resources such as user manuals, FAQs, and online tutorials.

    - **Data Security and Integrity**: Ensure that the e-procurement system has robust data security measures in place to protect sensitive procurement information. Implement access controls, encryption, and regular security audits to safeguard the system.

- **Expected Outcomes**:
    - Enhanced transparency and accountability in procurement processes.
    - Increased efficiency and reduced administrative costs in procurement activities.
    - Improved accessibility and participation of suppliers in procurement opportunities.

Strengthen Monitoring and Oversight Mechanisms

- **Actions**:
    - **Independent Procurement Oversight Body**: Establish or strengthen an independent body to oversee and audit public procurement activities. This body should have the authority to investigate procurement irregularities and enforce compliance with procurement regulations.
    - **Regular Procurement Audits**: Conduct regular audits of procurement activities to ensure compliance with established guidelines and procedures. Publish audit findings and take corrective actions as necessary.
    - **Real-Time Monitoring Systems**: Implement real-time monitoring systems to track procurement activities and identify potential issues early.

Use data analytics to monitor procurement trends and detect anomalies.

- **Whistleblower Protections**: Enhance protections for whistleblowers who report procurement fraud and misconduct. Ensure that whistleblower reports are investigated thoroughly and that whistleblowers are protected from retaliation.

- Expected Outcomes:
    - Increased accountability and reduced corruption in procurement processes.
    - Enhanced compliance with procurement regulations and guidelines.
    - Improved detection and prevention of procurement irregularities.

Promote Competition and Encourage Broad Participation

- Actions:
    - **Open and Competitive Bidding Processes**: Ensure that all procurement opportunities are advertised widely and conducted through open and competitive bidding processes. This promotes fairness and maximizes value for money.

- **Supplier Diversity Programs**: Implement programs to encourage the participation of diverse suppliers, including small and medium-sized enterprises (SMEs), women-owned businesses, and local firms. Provide training and support to help these suppliers compete effectively.

- **Supplier Registration and Prequalification**: Develop a streamlined process for supplier registration and prequalification. Maintain a comprehensive database of prequalified suppliers to ensure a broad pool of potential bidders.

- **Feedback and Grievance Mechanisms**: Establish clear mechanisms for suppliers to provide feedback and lodge grievances related to procurement processes. Ensure that these mechanisms are transparent and responsive.

- **Expected Outcomes**:

  - Increased competition and participation in public procurement.

  - Enhanced value for money in procurement contracts.

  - Improved supplier diversity and support for local businesses.

Enhance Capacity and Professionalism in Procurement

- Actions:

    - **Professional Development Programs**: Implement continuous professional development programs for procurement officials. Focus on building skills in procurement planning, contract management, negotiation, and ethical standards.

    - **Certification and Accreditation**: Develop certification and accreditation programs for procurement professionals. Encourage continuous learning and adherence to professional standards.

    - **Knowledge Sharing and Best Practices**: Foster a culture of knowledge sharing and continuous improvement within the procurement community. Establish forums, workshops, and online platforms for procurement professionals to share best practices and innovations.

    - **Collaboration with International Organizations**: Partner with international procurement organizations to access resources, expertise, and best practices. Participate in global procurement networks and forums to stay updated on emerging trends and innovations.

- **Expected Outcomes**:
    - Enhanced capacity and professionalism within the procurement workforce.
    - Improved procurement practices and outcomes.
    - Greater alignment with international procurement standards and best practices.

Conclusion: By implementing these detailed actions, the Public Procurement Authority can significantly enhance the fairness, transparency, and efficiency of public procurement processes in Sierra Leone. Developing and enforcing transparent procurement guidelines, implementing an e-procurement system, strengthening monitoring mechanisms, promoting competition, and enhancing capacity and professionalism are critical steps. These efforts will lead to increased consistency and transparency, reduced procurement fraud, improved efficiency, and greater participation of diverse suppliers. Ultimately, this will contribute to better value for money, reduced corruption, and increased public trust in government procurement activities, supporting sustainable development and good governance in Sierra Leone.

16. Ombudsman Office

- **Focus**: Address public grievances and ensure accountability.

- **Implementation**:

Provide Accessible Channels for Lodging Complaints

  - Actions:
    - **Multiple Access Points**: Establish multiple access points for lodging complaints, including walk-in offices, postal mail, telephone hotlines, email, and online portals. Ensure these access points are well-publicized and easily accessible to all citizens.

    - **User-Friendly Online Portal**: Develop a user-friendly online portal where citizens can submit complaints, track their status, and receive updates. Ensure the portal is accessible to people with disabilities and available in multiple languages.

    - **Community Outreach**: Conduct outreach programs to educate the public about the Ombudsman Office and how to lodge complaints. Use community meetings, workshops, and media campaigns to reach diverse audiences.

- **Mobile Complaint Units**: Establish mobile compliant units to reach remote and underserved areas. These units can visit communities on a scheduled basis to collect complaints and provide information about the Ombudsman's services.

- Expected Outcomes:
    - Increased accessibility and convenience for citizens to lodge complaints.
    - Greater awareness of the Ombudsman Office and its services.
    - Enhanced public trust in the grievance redressal system.

Ensure Prompt and Fair Investigation of Complaints

- Actions:
    - **Standardized Procedures**: Develop and implement standardized procedures for receiving, acknowledging, investigating, and resolving complaints. Ensure these procedures are transparent and adhere to principles of fairness and impartiality.
    - **Timely Resolution**: Set clear timelines for the investigation and resolution of complaints. Monitor compliance with these timelines to ensure timely redressal of grievances.

- **Qualified Investigators**: Employ and train a team of qualified investigators to handle complaints. Provide ongoing training on investigative techniques, legal standards, and customer service.

- **Collaboration with Agencies**: Establish protocols for collaborating with other government agencies and organizations to investigate complaints effectively. Ensure that there is a clear process for sharing information and coordinating actions.

- **Expected Outcomes**:

  - More efficient and effective handling of complaints.

  - Increased public satisfaction with the complaint resolution process.

  - Improved accountability and responsiveness of government agencies.

Regularly Review and Resolve Public Grievances Transparently

- **Actions**:

  - **Public Reporting**: Publish regular reports on the number, nature, and resolution of complaints received. Ensure these reports are accessible to the public and provide detailed information on the Ombudsman Office's activities and outcomes.

- **Public Hearings**: Conduct public hearings on significant or recurring issues identified through complaints. Use these hearings to gather additional information, engage with stakeholders, and develop recommendations for systemic improvements.

- **Case Studies and Best Practices**: Share case studies and best practices from resolved complaints to educate government agencies and the public on effective grievance redressal. Highlight successful resolutions and lessons learned.

- **Expected Outcomes**:

  - Increased transparency in the complaint resolution process.

  - Enhanced public confidence in the effectiveness of the Ombudsman Office.

  - Greater awareness and understanding of common issues and solutions.

Address Systemic Issues and Advocate for Policy Reforms

- **Actions**:

  - **Identify Systemic Issues**: Analyze complaint data to identify systemic issues and trends. Use this analysis to develop recommendations for policy and procedural reforms aimed at addressing root causes.

- **Advocacy and Recommendations**: Advocate for policy changes and procedural improvements based on findings from complaint investigations. Work with government agencies, legislators, and civil society organizations to implement these recommendations.

- **Monitoring and Follow-Up**: Monitor the implementation of recommended reforms and follow up with relevant agencies to ensure compliance. Provide regular updates to the public on the status of these reforms.

- **Stakeholder Engagement**: Engage with stakeholders, including civil society organizations, advocacy groups, and the media, to raise awareness of systemic issues and build support for proposed reforms.

- **Expected Outcomes**:

    - Improved policies and procedures that address systemic issues and prevent future grievances.

    - Increased accountability and responsiveness of government agencies.

    - Stronger advocacy and support for meaningful reforms.

Strengthen Institutional Capacity

- Actions:

    - **Training and Professional Development**: Provide continuous training and professional development opportunities for Ombudsman Office staff. Focus on areas such as investigation techniques, conflict resolution, customer service, and legal standards.

    - **Resource Allocation**: Ensure the Ombudsman Office has adequate resources, including funding, personnel, and technology, to carry out its mandate effectively. Advocate for necessary budget allocations and support.

    - **Performance Monitoring**: Implement robust performance monitoring systems to assess the effectiveness of the Ombudsman Office. Use performance data to inform strategic planning and continuous improvement efforts.

    - **Collaboration and Partnerships**: Foster collaboration with national and international ombudsman offices, human rights organizations, and other relevant bodies. Participate in networks and forums to share best practices and stay updated on emerging trends and innovations.

- **Expected Outcomes**:

    - Enhanced capacity and effectiveness of the Ombudsman Office.

    - Improved quality of service and complaint resolution.

    - Greater ability to address complex and systemic issues.

Conclusion: By implementing these detailed actions, the Ombudsman Office can significantly enhance its role in addressing public grievances and ensuring accountability in Sierra Leone. Providing accessible channels for lodging complaints, ensuring prompt and fair investigations, and regularly reviewing and resolving grievances transparently will improve public trust. Addressing systemic issues and advocating for policy reforms will lead to more effective governance. Strengthening institutional capacity will enhance the quality of service and complaint resolution. These efforts will contribute to a more transparent, responsive, and trustworthy government, fostering public trust and promoting good governance, thereby ensuring citizens' concerns are effectively addressed.

17. National Commission for Democracy

- **Focus**: Promote democratic education and public awareness.

- **Implementation**:

Promote Democratic Education

- Actions:

    - **Curriculum Development**: Collaborate with educational experts to develop a comprehensive democratic education curriculum for schools and universities. This curriculum should cover the principles of democracy, the importance of civic participation, human rights, and the roles and responsibilities of citizens in a democratic society.

    - **Teacher Training Programs**: Implement teacher training programs to equip educators with the knowledge and skills needed to teach democratic values and practices effectively. Provide ongoing professional development opportunities to ensure teachers remain updated on best practices and emerging issues in democratic education.

- **Civic Education Materials**: Produce and distribute high-quality educational materials, including textbooks, workbooks, multimedia resources, and interactive tools. Ensure these materials are accessible in multiple languages and formats to reach diverse student populations.

- **School Democracy Clubs**: Establish democracy clubs in schools to engage students in practical activities that promote democratic values. These clubs can organize debates, mock elections, community service projects, and discussions on current events.

- **Expected Outcomes**:

    - Increased awareness and understanding of democratic principles among students.

    - Enhanced ability of teachers to deliver effective democratic education.

    - Greater student engagement in democratic practices and civic participation.

Conduct Public Awareness Campaigns

- Actions:

    - **Media Campaigns**: Launch multimedia campaigns to raise public awareness about democracy, civic rights, and responsibilities. Use television, radio, social media, and print media to reach a wide audience. Highlight the importance of democratic participation and the impact of informed citizenry.

    - **Community Outreach Programs**: Organize community outreach programs to educate citizens about democratic values and practices. Hold workshops, town hall meetings, and public forums to engage directly with communities and address their specific needs and concerns.

    - **Democracy Festivals and Events**: Host annual democracy festivals and events to celebrate democratic achievements and promote civic engagement. These events can include speeches, panel discussions, cultural performances, and exhibitions that highlight the significance of democracy.

    - **Educational Campaigns on Elections**: Conduct targeted campaigns to educate the public about the electoral process, including voter registration, voting procedures, and the importance of participating in elections.

Partner with the National Electoral Commission to ensure accurate and timely information dissemination.

- **Expected Outcomes:**

    - Increased public awareness and appreciation of democratic values.

    - Higher levels of civic participation and voter turnout.

    - Strengthened public commitment to democratic principles and processes.

Facilitate Public Forums for Dialogue on Democratic Governance

- **Actions:**

    - **National and Regional Dialogues:** Organize national and regional dialogues to provide platforms for citizens to discuss democratic governance issues. Ensure these dialogues are inclusive and represent diverse perspectives, including those of marginalized and underrepresented groups.

    - **Policy Roundtables:** Convene policy roundtables that bring together government officials, civil society organizations, academics, and citizens to discuss and develop recommendations on key democratic governance issues.

Use these discussions to inform policy development and reforms.

- **Online Discussion Platforms**: Create online platforms for citizens to engage in discussions about democracy and governance. Use social media, forums, and webinars to facilitate these conversations and ensure broad participation.

- **Collaborations with Civil Society**: Partner with civil society organizations to co-host forums and dialogues on democratic governance. Leverage their networks and expertise to reach a wider audience and enrich the discussions.

- **Expected Outcomes**:

  - Enhanced public dialogue and engagement on democratic governance issues.

  - Increased influence of citizen input on policy decisions.

  - Strengthened collaboration between government and civil society.

Monitor and Advocate for Democratic Reforms

- Actions:

    - **Democratic Governance Assessments**: Conduct regular assessments of the state of democracy in Sierra Leone. Use indicators such as political freedoms, civil liberties, and citizen participation to evaluate democratic progress and identify areas for improvement.

    - **Advocacy for Reforms**: Advocate for democratic reforms based on the findings of governance assessments and public consultations. Engage with policymakers, legislators, and government officials to promote the adoption of recommended reforms.

    - **Public Reporting**: Publish annual reports on the state of democracy in Sierra Leone. Include data, analysis, and recommendations for strengthening democratic governance. Ensure these reports are widely disseminated and accessible to the public.

    - **Watchdog Role**: Act as a watchdog to monitor government actions and policies that affect democratic governance.

Report any threats to democracy, such as corruption, human rights violations, or restrictions on freedoms, and mobilize public and international support to address these issues.

- **Expected Outcomes**:

    - Improved policies and practices that support democratic governance.

    - Greater accountability of government actions and policies.

    - Increased public trust and confidence in democratic institutions.

Build Capacity for Democratic Governance

- **Actions**:

    - **Training Programs for Government Officials**: Provide training programs for government officials on democratic governance principles and practices. Focus on areas such as transparency, accountability, public participation, and human rights.

    - **Capacity Building for Civil Society**: Strengthen the capacity of civil society organizations to engage in democratic governance.

Provide training, resources, and support to enhance their advocacy, monitoring, and public engagement activities.

- **Research and Knowledge Sharing**: Conduct research on democratic governance issues and share findings with stakeholders. Create platforms for knowledge exchange, such as conferences, workshops, and publications.

- **International Partnerships**: Establish partnerships with international organizations and democratic institutions to share best practices, access technical assistance, and collaborate on democratic governance initiatives.

- **Expected Outcomes**:

    - Enhanced capacity of government officials and civil society to promote and uphold democratic governance.

    - Increased knowledge and application of democratic principles in governance.

    - Stronger international cooperation and support for democratic development.

Conclusion: By implementing these detailed actions, the National Commission for Democracy can significantly enhance the promotion of democratic education, public awareness, and governance in Sierra Leone. Promoting democratic education through curriculum development, teacher training, and school clubs will increase awareness among students. Conducting public awareness campaigns, facilitating public forums, and monitoring democratic reforms will strengthen civic participation and public trust in democracy. Building capacity for democratic governance among government officials and civil society will ensure sustainable democratic practices. These efforts will contribute to a more informed, engaged, and participatory citizenry, fostering a robust and resilient democracy in Sierra Leone.

18. Ministry of Planning and Economic Development

- **Focus**: Develop inclusive and participatory planning processes.

- **Implementation**:

  Ensure Planning Processes Involve Diverse Community Input

    - Actions:
        - **Stakeholder Consultations**: Conduct extensive stakeholder consultations during the planning process.

This includes engaging with local communities, civil society organizations, private sector representatives, and marginalized groups to gather their input and perspectives.

- **Public Hearings and Forums**: Organize public hearings and forums to discuss proposed development plans and policies. Ensure these events are accessible and inclusive, provide translation services, and accommodate people with disabilities.

- **Participatory Workshops**: Hold participatory workshops to involve citizens directly in the planning process. Methods such as participatory rural appraisal (PRA) and community mapping should be used to ensure active engagement and input from all community members.

- **Feedback Mechanisms**: Implement feedback mechanisms, such as online surveys, suggestion boxes, and community liaisons, to gather ongoing input from the public. Ensure that feedback is reviewed and incorporated into planning decisions.

- **Expected Outcomes**:
    - Increased inclusivity and representation in the planning process.

- Development plans that better reflect the needs and priorities of diverse communities.
- Enhanced public trust and ownership of development initiatives.

Develop Policies That Promote Equitable and Inclusive Development

- Actions:
    - **Equity Assessments**: Conduct equity assessments to identify and address disparities in access to resources, opportunities, and services. Use the findings to inform policy development and ensure that all groups benefit equitably from development initiatives.
    - **Targeted Programs for Vulnerable Groups**: Develop and implement targeted programs for vulnerable and marginalized groups, such as women, youth, persons with disabilities, and rural communities. Ensure these programs address specific barriers and promote inclusive development.
    - **Resource Allocation**: Ensure equitable allocation of resources in development projects. Prioritize investments in underserved and disadvantaged areas to reduce regional disparities and promote balanced development.

- **Inclusive Growth Strategies**: Formulate inclusive growth strategies that promote job creation, economic diversification, and sustainable development. Focus on sectors that have the potential to benefit the most vulnerable populations, such as agriculture, small and medium enterprises (SMEs), and renewable energy.

- **Expected Outcomes**:

    - Reduced disparities in access to resources and opportunities.

    - Improved socio-economic conditions for vulnerable and marginalized groups.

    - More balanced and inclusive economic growth.

Strengthen Institutional Capacity for Effective Planning and Implementation

- Actions:

    - **Capacity Building Programs**: Implement capacity building programs for planning and development officials. Focus on areas such as data analysis, project management, participatory planning, and policy development.

- **Technical Assistance and Training**: Provide technical assistance and training to local government units to enhance their planning and implementation capabilities. Partner with international organizations and development agencies to access expertise and resources.

- **Integrated Planning Systems**: Develop and implement integrated planning systems that facilitate coordination between different levels of government and sectors. Use digital tools and platforms to streamline planning processes and improve data sharing.

- **Monitoring and Evaluation Frameworks**: Establish robust monitoring and evaluation frameworks to track the progress and impact of development plans and policies. Use the findings to inform continuous improvement and ensure accountability.

- **Expected Outcomes**:

    - Enhanced capacity of government officials to develop and implement effective development plans.

    - Improved coordination and collaboration between different government agencies and sectors.

- Greater transparency and accountability in the planning and implementation process.

Engage in Strategic Planning for Long-Term Development

- Actions:
    - **National Development Plan**: Develop a comprehensive National Development Plan that outlines long-term development goals, strategies, and priorities. Ensure the plan is based on thorough research, data analysis, and stakeholder input.
    - **Vision and Goals Setting**: Establish a clear vision and specific, measurable goals for the country's development. Ensure these goals align with international development frameworks, such as the Sustainable Development Goals (SDGs).
    - **Scenario Planning and Risk Analysis**: Conduct scenario planning and risk analysis to anticipate potential challenges and uncertainties. Develop contingency plans and strategies to mitigate risks and ensure resilience.
    - **Regular Plan Reviews**: Conduct regular reviews and updates of the National Development Plan to ensure it remains relevant and responsive to changing circumstances.

Involve stakeholders in the review process to ensure continuous improvement.

- **Expected Outcomes**:
    - Clear and coherent long-term development strategy for Sierra Leone.
    - Increased resilience and adaptability to changing conditions and challenges.
    - Enhanced ability to achieve sustainable development goals.

Foster Partnerships and Collaboration

- **Actions**:
    - **Public-Private Partnerships (PPPs)**: Promote and facilitate public-private partnerships to leverage private sector resources and expertise for development projects. Develop clear guidelines and frameworks for PPPs to ensure transparency and mutual benefits.
    - **International Cooperation**: Strengthen international cooperation and partnerships with development agencies, donor organizations, and foreign governments. Engage in knowledge sharing, technical assistance, and joint initiatives to support national development goals.

- **Civil Society Engagement**: Collaborate with civil society organizations to enhance public participation, advocacy, and oversight in the development process. Provide support and resources to civil society groups to strengthen their capacity and effectiveness.

- **Academic and Research Institutions**: Partner with academic and research institutions to support evidence-based planning and policy development. Encourage research and innovation to address development challenges and promote sustainable solutions.

- **Expected Outcomes**:

  - Increased investment and resource mobilization for development projects.

  - Enhanced collaboration and synergy between government, private sector, and civil society.

  - Improved access to international expertise, resources, and best practices.

Conclusion: By implementing these detailed actions, the Ministry of Planning and Economic Development can significantly enhance the inclusivity, effectiveness, and sustainability of development planning in Sierra Leone.

Ensuring diverse community input, developing equitable policies, and strengthening institutional capacity will create development plans that reflect the needs of all citizens. Engaging in strategic planning and fostering partnerships will support long-term growth and resilience. These efforts will contribute to a more equitable and prosperous society, fostering long-term economic growth and social well-being. Through inclusive and participatory planning processes, Sierra Leone can achieve sustainable development and improve the quality of life for all its citizens.

19. Ministry of Healthcare and Welfare

Focus: Enhance the health and well-being of all citizens through accessible, quality healthcare services and comprehensive welfare programs.

Strengthen Healthcare Infrastructure and Services:

Actions:

- **Expand Healthcare Facilities:**
    - Construct and renovate hospitals, clinics, and health centers across the country, prioritizing underserved rural areas.
    - Ensure all healthcare facilities are equipped with essential medical supplies and modern technology.

- **Develop Health Workforce:**
    - Increase the recruitment and training of healthcare professionals, including doctors, nurses, and community health workers.
    - Implement continuous professional development programs to keep healthcare workers updated on best practices and new technologies.

- **Enhance Primary Healthcare:**
    - Strengthen primary healthcare services to provide preventive, curative, and rehabilitative care.
    - Promote community-based healthcare initiatives to improve local health outcomes.

Expected Outcomes:

- Improved access to quality healthcare services for all citizens.
- Enhanced capacity of healthcare facilities to meet the growing health needs.
- Increased number of trained healthcare professionals, leading to better health service delivery.

Promote Public Health and Preventive Care:

Actions:

- **Public Health Campaigns:**

    - Launch nationwide public health campaigns focused on disease prevention, vaccination, hygiene, and healthy lifestyles.

    - Utilize various media channels to disseminate information on health promotion and disease prevention.

- **Immunization Programs:**

    - Increase immunization coverage to protect against common and preventable diseases, particularly among children.

    - Ensure the availability of vaccines and streamline the distribution process to reach remote areas.

- **Nutrition Programs:**

    - Implement nutrition programs targeting vulnerable populations, including pregnant women, infants, and children.

    - Promote balanced diets and nutritional education to combat malnutrition and related health issues.

Expected Outcomes:

- Reduced incidence of preventable diseases and improved overall public health.

- Increased immunization rates that lead to lower morbidity and mortality from vaccine-preventable diseases.

- Enhanced nutritional status of vulnerable groups, contributing to better health outcomes.

Strengthen Social Welfare Programs:

Actions:

- **Support for Vulnerable Populations:**

    - Develop comprehensive social welfare programs for orphans, the elderly, people with disabilities, and other vulnerable groups.

    - Provide financial assistance, healthcare, and housing support to those in need.

- **Child Welfare Services:**

    - Strengthen child protection services to prevent abuse, neglect, and exploitation.

    - Implement programs to support children's education, health, and well-being, especially those from disadvantaged backgrounds.

- **Elderly Care Programs:**
    - Establish elderly care centers and community support programs to ensure the well-being of older adults.
    - Provide health services, social activities, and financial support to improve the quality of life for the elderly.

Expected Outcomes:

- Improved living conditions and well-being for vulnerable populations.
- Enhanced child protection and support systems, leading to better outcomes for children.
- Increased support and care for the elderly, promoting dignity and quality of life in their later years.

Technology Integration:

Actions:

- **Electronic Health Records (EHR):**
    - Implement EHR systems to streamline patient information management and improve healthcare delivery.
    - Ensure healthcare professionals are trained to use EHR systems effectively.

- **Telemedicine Services:**

    - Develop telemedicine programs to provide remote consultations and medical advice, especially in rural and remote areas.

    - Equip healthcare facilities with the necessary technology to support telemedicine.

Expected Outcomes:

- Improved efficiency in healthcare delivery through better management of patient information.

- Increased access to healthcare services for remote and underserved populations via telemedicine.

Training and Capacity Building:

Actions:

- **Professional Development Programs:**

    - Conduct regular training programs for healthcare professionals on topics such as patient care, medical ethics, and new medical technologies.

    - Partner with international health organizations to provide advanced training and exchange programs for healthcare workers.

- **Community Health Training:**

    - Train community health workers to provide basic healthcare services and health education at the grassroots level.

- Develop programs to raise awareness about common health issues and preventive measures.

Expected Outcomes:

- Enhanced knowledge and skills among healthcare professionals, leading to improved patient care.

- Empowered community health workers capable of addressing local health needs and promoting health education.

Conclusion: By implementing these comprehensive actions, the Ministry of Healthcare and Welfare can significantly enhance the health and well-being of all Sierra Leoneans. Strengthening healthcare infrastructure, promoting public health, integrating advanced technologies, and providing continuous training for healthcare professionals will improve access to and quality of healthcare services. Additionally, robust social welfare programs will support vulnerable populations, ensuring that no one is left behind. These efforts will contribute to a healthier, more equitable society, fostering overall socio-economic development and enhancing the quality of life for every citizen.

20. Ministry of Transportation

Focus: Develop a safe, efficient, and sustainable transportation network to enhance connectivity and support economic growth.

Develop Comprehensive Transportation Infrastructure:

Actions:

- **Road and Highway Construction:**
    - Construct and maintain roads, highways, and bridges to improve connectivity across the country.
    - Prioritize infrastructure projects that connect rural areas to urban centers, facilitating trade and access to services.

- **Public Transportation Systems:**
    - Expand and modernize public transportation systems, including buses, railways, and ferries.
    - Implement policies to improve the affordability, reliability, and safety of public transport.

Expected Outcomes:

- Enhanced national connectivity through improved road and transportation networks.
- Increased accessibility to public transportation, reducing travel time and costs for citizens.

Promote Road Safety:

Actions:

- **Traffic Law Enforcement:**

    - Strengthen the enforcement of traffic laws and regulations to reduce road accidents and improve safety.

    - Implement regular road safety campaigns to educate drivers and pedestrians on safe road usage.

- **Infrastructure Improvements:**

    - Develop pedestrian-friendly infrastructure, including sidewalks, crosswalks, and traffic signals.

    - Install road safety features such as speed bumps, guardrails, and reflective signage.

Expected Outcomes:

- Reduced road traffic accidents and fatalities.

- Improved safety for pedestrians and drivers, leading to safer road environments.

Enhance Sustainable Transportation:

Actions:

- **Environmentally Friendly Transport:**

    - Promote the use of electric vehicles (EVs) through incentives and the development of EV charging infrastructure.

    - Encourage non-motorized transport options, such as cycling and walking, by developing dedicated lanes and pathways.

- **Public Transport Upgrades:**

    - Invest in eco-friendly public transportation options, including electric buses and trains.

    - Implement policies to reduce emissions and promote energy-efficient transport solutions.

Expected Outcomes:

- Increased adoption of environmentally friendly transportation methods.

- Reduced carbon emissions and improved air quality through sustainable transport initiatives.

Improve Urban Transportation:

Actions:

- **Urban Transport Planning:**

    - Develop integrated urban transport plans to address congestion and improve mobility in cities.

    - Implement measures to improve public transportation services, such as dedicated bus lanes and frequent service schedules.

- **Smart City Solutions:**

    - Utilize smart technologies to manage traffic flow, monitor transport systems, and provide real-time information to commuters.

    - Develop mobile apps and online platforms to facilitate easy access to public transportation information.

Expected Outcomes:

- Reduced urban traffic congestion and improved mobility.

- Enhanced commuter experience through better-managed and accessible urban transport systems.

Strengthen Rural Connectivity:

Actions:

- **Rural Road Development:**

    - Enhance rural transport infrastructure to improve access to markets, healthcare, education, and other essential services.

    - Develop feeder roads and ensure regular maintenance to keep rural areas connected.

- **Transport Services Expansion:**

    - Increase the availability of public transport services in rural areas, including buses and minibusses.

    - Provide subsidies or incentives to encourage transport providers to serve remote areas.

Expected Outcomes:

- Improved access to essential services and economic opportunities for rural populations.

- Enhanced rural-urban connectivity, promoting balanced regional development.

Training and Capacity Building:

Actions:

- **Professional Development Programs:**
    - Implement training programs for transportation sector workers, including engineers, planners, and maintenance staff.
    - Establish partnerships with international transportation organizations for knowledge exchange and capacity building.

- **Community Engagement:**
    - Involve local communities in transportation planning and decision-making processes to ensure services meet their needs.
    - Conduct public awareness campaigns on road safety and sustainable transportation practices.

Expected Outcomes:

- Enhanced skills and knowledge among transportation sector professionals.
- Increased community involvement and satisfaction with transportation services.

Conclusion: The proposed actions for the Ministry of Transportation aim to develop a safe, efficient, and sustainable transportation network that supports economic growth and enhances connectivity across Sierra Leone. By focusing on infrastructure development, road safety, sustainable transportation, and improved urban and rural connectivity, the ministry can significantly improve the mobility and accessibility of goods and people. These initiatives will not only reduce travel time and costs but also promote environmental sustainability and safety. Through continuous professional development and community engagement, the transportation sector will become more responsive to the needs of the populace, contributing to the nation's overall development and prosperity.

21. Ministry of Energy and Power

Focus: Ensure reliable, affordable, and sustainable energy supply to support economic growth and improve the quality of life for all citizens.

Expand Energy Infrastructure:

Actions:

- **Develop Power Generation Capacity:**
    - Invest in building new power plants and upgrading existing facilities, focusing on renewable energy sources such as hydro, solar, and wind.
    - Implement measures to reduce energy losses in transmission and distribution networks.

- **Rural Electrification:**
  - Extend the national grid to rural and underserved areas to provide reliable electricity.
  - Promote off-grid renewable energy solutions in remote regions, such as solar home systems and mini-grids.

Expected Outcomes:

- Increased power generation capacity and reduced energy deficits.
- Enhanced access to electricity in rural areas, improving quality of life and economic opportunities.

Promote Renewable Energy:

Actions:

- **Renewable Energy Policies:**
  - Develop and implement policies that incentivize the adoption of renewable energy technologies.
  - Provide subsidies and financial incentives for households and businesses to invest in solar panels, wind turbines, and other renewable energy systems.

- **Public Awareness Campaigns:**
  - Launch campaigns to educate the public on the benefits of renewable energy and energy efficiency.

- Promote the adoption of energy-saving practices in homes and businesses.

Expected Outcomes:

- Increased use of renewable energy, reducing reliance on fossil fuels.
- Enhanced public awareness and participation in sustainable energy practices.

Improve Energy Efficiency:

Actions:

- **Energy Efficiency Standards:**
    - Establish and enforce energy efficiency standards for appliances, buildings, and industrial processes.
    - Implement energy audit programs for businesses and public institutions to identify and address inefficiencies.
- **Modernize Grid Infrastructure:**
    - Upgrade the national grid to reduce energy losses and improve reliability.
    - Implement smart grid technologies to enhance grid management and efficiency.

Expected Outcomes:

- Reduced energy consumption and lower energy costs for consumers and businesses.

- Enhanced reliability and efficiency of the national grid.

Strengthen Institutional Capacity:

Actions:

- **Professional Development:**
    - Conduct training programs for energy sector professionals on the latest technologies and best practices.
    - Establish partnerships with international energy organizations for knowledge exchange and capacity building.

- **Regulatory Framework:**
    - Strengthen the regulatory framework to support investment in the energy sector.
    - Ensure transparent and fair regulations to attract private sector participation.

Expected Outcomes:

- Improved skills and knowledge among energy sector professionals.
- Enhanced regulatory environment that supports sustainable energy development.

Conclusion: By executing the proposed actions, the Ministry of Energy and Power can significantly improve the reliability, affordability, and sustainability of the energy supply in Sierra Leone. Expanding energy infrastructure, promoting renewable energy, enhancing energy efficiency, and strengthening institutional capacity will lead to increased power generation and access, particularly in rural areas. These efforts will not only support economic growth and development but also improve the quality of life for all citizens. Through the adoption of sustainable practices and technologies, the ministry will contribute to a greener and more resilient energy sector, fostering a sustainable future for Sierra Leone.

22. Ministry of Fisheries and Wildlife

Focus: Sustainably manage and protect fisheries and wildlife resources to support biodiversity, economic development, and community livelihoods.

Sustainable Fisheries Management:

Actions:

- **Fisheries Regulation:**
    - Implement and enforce regulations to prevent overfishing and ensure sustainable fish stocks.
    - Establish quotas, seasonal restrictions, and protected areas to manage fish populations effectively.

- **Support for Fishing Communities:**

    - Provide training and resources to fishing communities on sustainable fishing practices.

    - Develop programs to improve the economic resilience of fishing communities, such as alternative livelihood projects.

Expected Outcomes:

- Sustainable fish stocks and healthier marine ecosystems.

- Improved livelihoods and economic stability for fishing communities.

Wildlife Conservation:

Actions:

- **Protected Areas:**

    - Expand and manage protected areas to conserve critical wildlife habitats.

    - Implement anti-poaching measures and enhance enforcement to protect endangered species.

- **Community Involvement:**

    - Engage local communities in wildlife conservation efforts through education and incentive programs.

    - Promote eco-tourism as a sustainable way to generate income and support conservation.

Expected Outcomes:

- Increased protection and preservation of wildlife and their habitats.
- Enhanced community participation and support for wildlife conservation.

Biodiversity Protection:

Actions:

- **Habitat Restoration:**
    - Implement programs to restore degraded habitats and ecosystems.
    - Promote reforestation and sustainable land management practices.
- **Research and Monitoring:**
    - Conduct research to monitor wildlife populations and biodiversity health.
    - Use data to inform conservation strategies and policy decisions.

Expected Outcomes:

- A thriving and healthy ecosystems that support diverse species.
- Improved data and insights to guide effective conservation efforts.

Strengthen Institutional Capacity:

Actions:

- **Training and Capacity Building:**
  - Provide training for fisheries and wildlife management staff on best practices and new technologies.
  - Establish partnerships with international conservation organizations for knowledge exchange and capacity building.

- **Policy and Legislation:**
  - Develop and implement policies and legislation that support sustainable fisheries and wildlife management.
  - Ensure regulatory frameworks are enforced effectively to protect resources.

Expected Outcomes:

- Enhanced skills and capacity among fisheries and wildlife management professionals.
- Stronger policy and legal frameworks that support sustainable resource management.

Conclusion: Implementing the detailed actions for the Ministry of Fisheries and Wildlife will ensure the sustainable management and protection of Sierra Leone's valuable natural resources. By regulating fisheries, conserving wildlife, protecting biodiversity, and strengthening institutional capacity, the ministry will promote ecological balance and enhance community livelihoods. These measures will support the health of marine and terrestrial ecosystems, ensuring the long-term sustainability of fish stocks and wildlife populations. Engaging local communities and fostering eco-tourism will generate economic benefits while reinforcing conservation efforts. Together, these initiatives will contribute to the overall environmental, economic, and social well-being of Sierra Leone.

23. Ministry of Telecommunications and Broadband Internet

Focus: Enhance the telecommunications infrastructure to provide reliable, high-speed broadband internet and wireless services, fostering digital inclusion and economic growth.

Expand Telecommunications Infrastructure:

Actions:

- **Broadband Network Expansion:** -
    - Invest in the development and expansion of broadband infrastructure, including fiber-optic networks, to provide high-speed internet access nationwide.

- Prioritize the extension of broadband networks to rural and underserved areas to ensure digital inclusion.

- **Wireless Network Improvement:**
    - Upgrade existing wireless networks to improve coverage, capacity, and quality of service.
    - Deploy advanced wireless technologies, such as 4G and 5G, to enhance mobile internet speed and connectivity.

Expected Outcomes:

- Increased access to high-speed internet and wireless services across the country.
- Improved connectivity in rural and underserved areas, reducing the digital divide.

Promote Digital Literacy and Inclusion:

Actions:

- **Digital Literacy Programs:**
    - Implement nationwide digital literacy programs to educate citizens on the use of the internet and digital technologies.
    - Partner with educational institutions to integrate digital literacy into the curriculum at all levels.

- **Community Internet Access:**
  - Establish public internet access points, such as community centers and libraries, to provide free or affordable internet access.
  - Provide training and support to help communities utilize digital tools and resources effectively.

Expected Outcomes:

- Increased digital literacy and skills among the population.
- Enhanced digital inclusion, enabling more citizens to participate in the digital economy.

Strengthen Regulatory and Policy Framework:

Actions:

- **Telecommunications Policy Development:**
  - Develop and implement policies that encourage investment in telecommunications infrastructure and services.
  - Ensure regulatory frameworks are transparent, fair, and conducive to competition and innovation.

- **Consumer Protection:**
  - Establish regulations to protect consumers' rights, ensuring quality of service and fair pricing.
  - Implement mechanisms for resolving consumer

complaints and disputes effectively.

Expected Outcomes:

- A robust regulatory environment that supports growth and innovation in the telecommunications sector.
- Increased consumer trust and satisfaction with telecommunications services.

Enhance Cybersecurity and Data Protection:

Actions:

- **Cybersecurity Measures:**
    - Develop and implement a national cybersecurity strategy to protect telecommunications infrastructure and users from cyber threats.
    - Provide training and resources to enhance the cybersecurity capabilities of telecom operators and internet service providers.

- **Data Protection Policies:**
    - Establish and enforce data protection regulations to safeguard users' personal information.
    - Promote best practices for data management and security among businesses and public institutions.

Expected Outcomes:

- Enhanced security and resilience of the telecommunications infrastructure.

- Improved data protection and privacy for internet users.

Foster Innovation and Digital Economy:

Actions:

- **Support for Startups and Entrepreneurs:**
    - Create programs and incentives to support tech startups and digital entrepreneurs.
    - Establish innovation hubs and incubators to provide resources, mentorship, and networking opportunities.

- **E-Government Services:**
    - Develop and expand e-government services to improve public service delivery and accessibility.
    - Promote the use of digital platforms for government transactions and interactions with citizens.

Expected Outcomes:

- A vibrant digital economy with increased opportunities for startups and entrepreneurs.
- Enhanced efficiency and accessibility of government services through digital platforms.

Conclusion: By implementing these comprehensive actions, the Ministry of Telecommunications and Broadband Internet can significantly enhance the telecommunications infrastructure in Sierra Leone. Expanding broadband and wireless networks, promoting digital literacy, strengthening regulatory frameworks, enhancing cybersecurity, and fostering innovation will ensure reliable, high-speed internet access for all citizens. These efforts will drive digital inclusion, economic growth, and social development, positioning Sierra Leone as a leader in the digital age and improving the quality of life for its citizens.

24. Chamber of Commerce and Business Affairs

Focus: Promote business development, advocate for economic policies, and support local enterprises to enhance economic growth and competitiveness.

Support Business Development:

Actions:

- **Business Advisory Services:**
    - Provide advisory services to businesses, including startups and SMEs, on topics such as business planning, marketing, finance, and compliance.
    - Offer mentorship programs to connect experienced business leaders with new entrepreneurs.

- **Access to Finance:**

    - Facilitate access to financial resources for businesses by partnering with financial institutions to offer loans, grants, and other funding opportunities.

    - Develop programs to improve businesses' financial literacy and management skills.

Expected Outcomes:

- Enhanced capacity and sustainability of local businesses.

- Increased access to financial resources, enabling business growth and innovation.

Advocate for Business-Friendly Policies:

Actions:

- **Policy Advocacy:**

    - Work with government agencies to develop and advocate for policies that support business growth and economic development.

    - Engage in regular dialogue with policymakers to address the concerns and needs of the business community.

- **Regulatory Reforms:**

    - Identify and advocate for reforms to reduce bureaucratic hurdles and improve the ease of doing business.

- Promote the establishment of transparent and efficient regulatory frameworks.

Expected Outcomes:

- A more business-friendly policy environment that encourages investment and entrepreneurship.
- Reduced regulatory barriers, making it easier for businesses to operate and expand.

Promote Trade and Investment:

Actions:

- **Trade Missions and Expos:**
  - Organize trade missions, fairs, and expos to promote local businesses and attract foreign investment.
  - Facilitate networking opportunities between local businesses and international investors.

- **Market Access:**
  - Support businesses in accessing new markets through export promotion programs and international trade partnerships.
  - Provide information and resources on international trade regulations and opportunities.

Expected Outcomes:

- Increased trade and investment flows, boosting economic growth.
- Enhanced market access for local businesses, leading to greater export opportunities.

Foster Innovation and Entrepreneurship:

Actions:

- **Innovation Hubs and Incubators:**
    - Establish innovation hubs and business incubators to support startups and foster a culture of entrepreneurship.
    - Provide resources, mentorship, and networking opportunities to help startups grow and succeed.
- **Entrepreneurship Programs:**
    - Develop programs to promote entrepreneurship, particularly among youth and women.
    - Offer training and workshops on entrepreneurship, business management, and innovation.

Expected Outcomes:

- A vibrant ecosystem for innovation and entrepreneurship.
- Increased number of successful startups and small businesses, contributing to economic diversification.

Enhance Business Services and Support:

Actions:

- **Networking and Collaboration:**

    - Organize networking events, forums, and conferences to facilitate collaboration and knowledge sharing among businesses.

    - Create platforms for businesses to engage with each other and with potential partners, customers, and investors.

- **Business Information Services:**

    - Provide up-to-date information on market trends, economic conditions, and business opportunities.

    - Develop online portals and resources to easily access business information and services.

Expected Outcomes:

- Strengthened business networks and increased collaboration.

- Improved access to critical information and resources for businesses.

Conclusion: While the Chamber of Commerce and Business Affairs is not a government entity, it plays a vital role in supporting and advocating for the business community in Sierra Leone. By providing advisory services, advocating for business-friendly policies, promoting trade and investment, fostering innovation, and enhancing business support services, the chamber can significantly contribute to the economic development and competitiveness of the country.

These efforts will help create a thriving business environment that supports entrepreneurship, attracts investment, and drives sustainable economic growth.

Case Study: Kampala Health, Agriculture and General Services Sierra Leone (KHAGES-SL) – Empowering Independence through Agriculture

Overview: Kampala Agricultural Farm, officially known as Kampala Health Agriculture & General Services (KHAGES-SL), is a pioneering initiative in Sierra Leone dedicated to empowering farmers, vulnerable women, marginalized men, orphans, and deprived children. By focusing on sustainable agriculture and community support, KHAGES-SL aims to combat hunger, reduce poverty, and promote economic independence. This case study explores how KHAGES-SL can serve as a model for fostering economic independence and empowerment in Sierra Leone, aligning with the objectives of the Chamber of Commerce and Business Affairs.

Mission and Objectives

- **Mission**: To collaborate with various organizations and groups to support orphans, farmers, vulnerable women, marginalized men, and deprived children.

- **Key Objective**: Empower farmers by providing seeds, seedlings, farm tools, food, infrastructure, and appropriate training to manage hunger and poverty effectively.

Support Business Development

Actions

1. **Business Advisory Services**:
   - KHAGES-SL offers advisory services to farmers and agricultural entrepreneurs on business planning, marketing, finance, and compliance. This support includes mentorship programs connecting experienced agricultural professionals with new farmers.

2. **Access to Finance**:
   - Facilitates access to financial resources by partnering with financial institutions to offer loans, grants, and other funding opportunities. Programs are developed to improve financial literacy and management skills among farmers.

Expected Outcomes

- Enhanced capacity and sustainability of local agricultural businesses.
- Increased access to financial resources, enabling business growth and innovation.

Advocate for Business-Friendly Policies

Actions

1. **Policy Advocacy**:
   - KHAGES-SL works with government agencies to develop and advocate for policies that support agricultural growth and economic development. This includes engaging in regular dialogue with policymakers to address the concerns and needs of the agricultural community.

2. **Regulatory Reforms**:
   - Identifies and advocates for reforms to reduce bureaucratic hurdles and improve the ease of doing business in the agricultural sector. Promotes the establishment of transparent and efficient regulatory frameworks.

Expected Outcomes

- A more business-friendly policy environment that encourages investment and entrepreneurship.
- Reduced regulatory barriers, making it easier for agricultural businesses to operate and expand.

Promote Trade and Investment

Actions

1. **Trade Missions and Expos**:
   - Organizes trade missions, fairs, and expos to promote local agricultural products and attract foreign investment. Facilitates networking opportunities between local farmers and international investors.

2. **Market Access**:
   - Supports farmers in accessing new markets through export promotion programs and international trade partnerships. Provides information and resources on international trade regulations and opportunities.

Expected Outcomes

- Increased trade and investment flows, boosting economic growth.
- Enhanced market access for local agricultural products, leading to greater export opportunities.

Foster Innovation and Entrepreneurship

Actions

1. **Innovation Hubs and Incubators**:
   - Establishes innovation hubs and business incubators to support agricultural startups and foster a culture of entrepreneurship. Provides resources, mentorship, and networking opportunities to help startups grow and succeed.

2. **Entrepreneurship Programs**:
   - Develops programs to promote entrepreneurship, particularly among youth and women. Offers training and workshops on entrepreneurship, business management, and innovation.

Expected Outcomes

- A vibrant ecosystem for innovation and entrepreneurship.

- Increased number of successful agricultural startups and small businesses, contributing to economic diversification.

Enhance Business Services and Support

Actions

1. **Networking and Collaboration**:
   - Organizes networking events, forums, and conferences to facilitate collaboration and knowledge sharing among agricultural businesses. Creates platforms for businesses to engage with each other and with potential partners, customers, and investors.

2. **Business Information Services**:
   - Provides up-to-date information on market trends, economic conditions, and business opportunities. Develops online portals and resources to offer easy access to business information and services.

Expected Outcomes

- Strengthened business networks and increased collaboration.
- Improved access to critical information and resources for agricultural businesses.

Conclusion: Kampala Agricultural Farm (KHAGES-SL) exemplifies how structured agricultural initiatives can foster economic independence and empowerment in Sierra Leone.

By providing comprehensive support to farmers, advocating for business-friendly policies, promoting trade and investment, fostering innovation, and enhancing business services, KHAGES-SL significantly contributes to economic development and competitiveness. This model can be expanded to allow more individuals to become independent income earners, ultimately empowering them to become owners of their political, social, and decision-making processes in life. This case study highlights the potential of agriculture as a pathway to economic independence and a brighter future for Sierra Leone.

Reflection and Engagement Questions

This chapter offers a detailed exploration of effective strategies to combat the detrimental principles that undermine democratic governance in Sierra Leone. It addresses the critical need to promote merit-based systems, enhance accountability and transparency, launch public awareness campaigns, advocate for legal and policy reforms, and support civil society organizations. By implementing these strategies, Sierra Leone can foster a political environment that upholds democratic values and promotes equitable development. The chapter aims to empower readers with practical approaches to counteract corruption, nepotism, political partisanship, and other harmful practices, ensuring a more stable and just society.

1. **Promoting Merit-Based Systems:**
   - How can the implementation of merit-based systems improve governance and reduce favoritism in Sierra Leone? Provide examples of successful merit-based initiatives.

   - _____
     _____

2. **Enhancing Accountability and Transparency:**
   - What measures can be taken to strengthen accountability and transparency in Sierra Leone's public sector? Reflect on the impact of these measures on public trust and governance efficiency.

   - _____
     _____

3. **Public Awareness Campaigns:**
   - How can public awareness campaigns be effectively utilized to educate citizens about their rights and responsibilities in combating corruption and other detrimental practices?

   - _____
     _____

4. **Advocating for Legal and Policy Reforms:**
   - What specific legal and policy reforms are necessary to address the root causes of governance issues in Sierra Leone? How can advocacy groups like SLAM influence these reforms?

   - _____
     _____

5. **Supporting Civil Society Organizations:**
   - In what ways can civil society organizations contribute to promoting good governance and democratic principles in Sierra Leone? Discuss the importance of collaboration between civil society and government institutions.

   - _____
     _____

These questions are designed to help readers reflect on the chapter's content, encouraging critical thinking and practical application of the strategies discussed to improve governance in Sierra Leone.

CHAPTER 10:
THE SIERRA LEONE ADVOCACY MOVEMENT

The Sierra Leone Advocacy Movement was established to mobilize the Sierra Leonean diaspora and foster a collaborative approach to national development. The organization aims to maximize Sierra Leone's potentials by promoting democratic principles, social justice, and economic development. SLAM believes in an open and democratic society where every citizen can develop and enjoy their talents without fear or undue influence.

The vision of SLAM is to forge a vibrant democratic Sierra Leone that champions unity, peace, and justice, as enshrined in the national coat of arms. This vision is reflected in their mission to identify local resources and potentials as a foundation for national and individual development. By leveraging the diaspora's resources and expertise, SLAM is committed to facilitating informed participation in the democratic process, ensuring that every Sierra Leonean can contribute to a society that values individuality, democratic engagement, and sustainable development.

SLAM's objectives include advocating for social justice, promoting the general welfare of Sierra Leoneans both at home and abroad, and fostering equality and gender parity.

The organization also focuses on building democratic institutions and mentalities by cooperating with global democracy promotion institutions.

As such, the history and evolution of governance in Sierra Leone, coupled with the ongoing challenges and efforts towards democratic consolidation, provide a rich context for understanding the current political landscape. By delving into these historical milestones and highlighting the role of organizations like SLAM, this book aims to offer a comprehensive view of Sierra Leone's journey towards good governance and democratic principles.

Mission: SLAM aims to mobilize Sierra Leone's private sector by identifying local resources and potentials as a foundation for national and individual development. We aim to cultivate an open and democratic society where citizens can develop and enjoy their talents without fear or undue influence. Through leveraging the diaspora's resources and expertise, we are committed to facilitating informed participation in the democratic process and ensuring that every Sierra Leonean can contribute to a society that values individuality, democratic engagement, and sustainable development.

Vision: Our vision is to forge a vibrant, democratic Sierra Leone that champions unity, peace, and justice, as enshrined in our national coat of arms. We are dedicated to building a society that celebrates individuality, empowers citizens through education to make informed choices for their futures—free from prejudice, and fosters robust private-sector participation. We envision a nation where every Sierra Leonean, regardless of their location, actively contributes to and thrives within a democratic framework that firmly rejects undemocratic values.

SLAM Constitution

Preamble

We, the members of the Sierra Leone Advocacy Movement, do hereby adopt this Constitution and swear to adhere to and execute, to the best of our abilities, the prescriptions herein outlined. This Initiative shall be non-political, non-ethnic, and secular.

CHAPTER I: NAME, VISION, MISSION AND VALUES

ARTICLE 1: NAME

The name of the Organisation shall be the Sierra Leone Advocacy Movement, hereinafter referred to as "SLAM' or "the Advocacy."

ARTICLE 2: VISION

Our vision is to forge a vibrant, democratic Sierra Leone that champions unity, peace, and justice, as enshrined in our national coat of arms. We are dedicated to building a society that celebrates individuality, empowers citizens through education to make informed choices for their futures free from prejudice, and fosters robust private-sector participation. We envision a nation where every Sierra Leonean, regardless of their location, actively contributes to and thrives within a democratic framework that firmly rejects undemocratic values.

ARTICLE 3: MISSION

SLAM aims to mobilise Sierra Leone's private sector by identifying local resources and potentials as a foundation for national and individual development. We aim to cultivate an open and democratic society where citizens can develop and enjoy their talents without fear or undue influence. Through leveraging the diaspora's resources and expertise, we are committed to facilitating informed participation in the democratic process, ensuring that every Sierra Leonean can contribute to a society that values individuality, democratic engagement, and sustainable development.

ARTICLE 4: CORE VALUES

1. Integrity:

1. We shall prioritise the highest quality in being honest, ethical, and transparent in all actions and decisions;
2. Honouring our commitments and promises to the communities we work in and the people we work with;
3. Be transparent in our dealings, including financial transactions, by providing regular updates on the use of funds and being open to questions and feedback from the communities; and
4. Adhering to ethical standards in our dealings, including avoiding conflicts of interest and actions that could harm the communities and the Initiative.

2. Community: We shall strive to:

1. Be inclusive of members of the communities, including those who may be marginalized;
2. Respect the communities we work in and the people we work with, including their cultures, traditions, and beliefs. This includes involving community members in decision-making and ensuring that their voices are heard and valued;
3. Work collaboratively with the communities to identify common goals and develop strategies to achieve them. This includes building partnerships with local organisations and stakeholders to leverage resources and expertise; and
4. Empower community members by providing them with the tools and resources they need to take control of their development. This includes providing training and capacity-building opportunities, as well as promoting local leadership and entrepreneurship.

3. Excellence: We shall explore new ideas, approaches, and technologies to deliver high-quality programs and services that meet the needs of our beneficiaries. This includes:

1. Setting clear standards for performance and outcomes and continuously monitoring and evaluating our programs to ensure that they are effective; and
2. Maintain a professional demeanor in all our interactions within and without the Initiative, including dressing appropriately, using appropriate language, and demonstrating respect and courtesy.

4. Accountability:

11. We shall and must be accountable to our beneficiaries, donors, and other stakeholders for our actions and decisions – this includes being transparent about the use of resources and outcomes achieved, as well as being willing to acknowledge and learn from mistakes.

CHAPTER II: OFFICES OF THE INITIATIVE

ARTICLE 5: PRINCIPAL AND ADDITIONAL OFFICES

1. The principal office of SLAM shall be situated in London, England, UK.

2. Other offices or chapters of the Initiative, apart from the main office, shall be opened and situated in relevant countries, regions, and localities for the purpose of fund raising and operations.

CHAPTER III: MANAGEMENT OF THE INITIATIVE

ARTICLE 6: FUNCTIONS AND SIZE OF BOARD

1. Function: All the powers of the SLAM shall be exercised by or under the authority of its Board of Directors or Executives (hereinafter called "Board/Executive"), and the activities and affairs of SLAM shall be managed by or under the direction, and subject to the oversight of the Board.

2. Size of Board: The Board shall consist of no less than ten (10) and no more than forty (40) persons (each a "Director/Executive"). The Board/Executive may, from time to time by action of the Board, increase the size of the current Board and elect additional Directors/Executives thereto or reduce the size of the current Board, provided that such deduction shall not shorten the term of any incumbent Directors/ Executives. The Director General (DG) shall serve as the chair of the Board of Directors/Executives with a voting right.

ARTICLE 7: TERM LIMITS

1. Each Director/Executive shall be elected for a term of up to five (5) years, with such term beginning on January 1 of the year following the year in which the Director/Executive is elected and ending on December 31 of the fifth year; or such other time that the Board deemed appropriate so long as it is within the meaning of a five year;

2. Directors/Executives shall be elected by action of the Board at its annual meeting. Except provided otherwise in this constitution or any other law, no Director may be elected to more than two (2) successive terms ("term limit");

3. A Director/Executive may be elected up to two (2) additional successive terms beyond the term limit ("extended term") if such Director/ Executives is also elected and continues to serve as a Board/ Executives Officer during such extended term; and

4. The Board may permit the election of a Director/ Executives beyond the Director's/ Executives' term limit or extended term if the Board determines that such action is in the best interest of SLAM.

ARTICLE 8: RESIGNATION, REMOVAL AND VACANCY

1. A Director/Executive may resign at any time by delivering a signed written notice to the Director General or Secretary, and such resignation shall be effective upon delivery of the notice unless the notice specifies a later effective date.

2. A Director/Executive may be removed from office by a majority vote of other Directors/Executives in office.

3. More than three (3) or more "unexcused absences" at Board meetings by any Director during any single term of office shall result in the Board taking a vote to remove such Director from the Board (with such removal requiring a majority vote of the Directors present at such meeting), either at the third Board meeting in which the Director is absent or, if elected by the Chair, the Co-Chair and the Vice Chairs in their sole joint discretion, at a subsequent Board meeting. "Unexcused absence" for the purpose of this paragraph means, a director's absence from a Board meeting without informing the Chair, President/CEO or Secretary of such absence prior to the commencement of such meeting.

4. A vacancy in any Director position may be filled for the balance of the term by action of the Board at any of its meetings.

ARTICLE 9: QUORUM AND VOTING OF THE BOARD

1. **Quorum**: At any meeting of the Board, one-third (1/3) of the entire number of Directors/ Executives in office shall constitute a quorum for the transaction of business.

A Director/ Executive will be considered present if such Director/ Executive attends the meeting either in person or by any means of communication (teleconference, videoconference, Zoom, WhatsApp, etc.) by which all Directors/ Executives participating in the meeting can hear each other simultaneously.

2. **Voting**: Each Director/Executive so present at any such meeting of the Board shall be counted for the purpose of voting. Except as required by any other law, policy, or Bylaws, the vote of a simple majority of the Directors/Executives present at a meeting at which a quorum is established shall constitute the action of the Board. A written record shall be made of the actions taken at such meeting and made a part of the records of the Board.

ARTICLE 10: MEETINGS OF THE BOARD

1. **Annual Meeting**: The annual meeting of the Board shall occur at the last month of the calendar year or at such time that the Board deems appropriate.

2. **Special Meetings**: Special meetings of the Board may be called by the Director General, the Secretary, or any three (3) Directors/Executives.

3. **Notice**: Notice of the date, time, and place of each meeting of the Board shall be given to each Director/Executive by the Secretary or the person or persons calling the meeting not less than ten (10) days before the date of such meeting. Such notice need not specify the purposes of the meeting.

The giving of notice shall be deemed to be waived by any Director/Executive who shall attend and participate in such meeting, unless the Director/Executive at the beginning of such meeting, or promptly upon the Director's/Executive's arrival there, objects to holding such meeting or transacting business at such meeting, and does not thereafter vote for or assent to any action taken at such meeting; or waives such notice requirement in a signed writing delivered by the Director/Executive to the Secretary either before, at or after such meeting.

CHAPTER IV: COMMITTEES OF THE BOARD; ADVISORY BODIES

ARTICLE 11. COMPOSITION OF EXECUTIVE COMMITTEE

The Executive Committee of the Board shall consist of the following members: Director General (DG); Deputy Director General (DDG), Secretary, Treasurer, and such other Director (s) as the Board may from time to time select by a majority vote of all Directors in office. However, provided that with respect to those members serving on the Executive Committee, such members' respective memberships on the Executive Committee shall be for no more than a maximum of five years, except that such limitation shall not apply to any of such members who during such memberships on the Executive Committee are elected to any of the Board officer positions identified above. In this case, such memberships on the Executive Committee, as applicable, shall be vacated and filled in accordance therewith. In addition to the members of the Executive Committee identified in the foregoing sentence, the Director General (DD) shall be a member and head of the Executive Committee.

ARTICLE 12: FUNCTION

The Executive Committee shall exercise the full authority of the Board in the management of the affairs of SLAM during the intervals between meetings of the Board, provided, however, that the Executive Committee shall not have the authority of the Board to:

1. Amend, restate, or repeal the Articles of this Constitution or any Bylaw or policy of the Initiative;
2. Elect or remove any Director or officer from the Board or from a Board committee;
3. Take, authorise, or ratify any action that violates the Articles of this Constitution or the Bylaws or policy of the Initiative or that is unlawful under applicable law;
4. Adopt a plan of merger or consolidation with another corporation or entity;
5. Authorise the sale, lease, exchange, or mortgage of all or substantially all of the Initiative's assets;
6. Authorise a voluntary dissolution of the Initiative or revoke proceedings thereof;
7. Amend, alter, or replace any resolution of the Board which, by its terms, provides that it shall not be amended, altered, or repealed by the Executive Committee; or
8. Remove the Director General (DG) and/or appoint a successor DG

ARTICLE 13: DEPUTY DIRECTOR GENERAL OF THE EXECUTIVE COMMITTEE

The Deputy General shall serve as the deputy of the Executive Committee, and the **DG** shall serve as the chair of the Executive Committee.

ARTICLE 14: MEETING QUORUM

At any meeting of the Executive Committee, the presence of at least a majority of the members of the Executive Committee shall be required to establish a quorum for the transaction of business. The vote of a majority of the members of the Executive Committee present at a meeting at which a quorum is established shall constitute the action of the Executive Committee.

ARTICLE 15: KEY DECISIONS

The Director General of the Executive Committee shall consult with the chairs of all other additional Board/Executive committees relevant to major actions to be taken by the Executive Committee between regular meetings of the Board. All major actions taken by the Executive Committee between meetings of the Board shall be reported to the Board at its next regular meeting, and all significant issues for discussion relevant to the Board shall be presented to the Board for review and consideration at such next regular meeting.

ARTICLE 16: OFFICERS

1. The Director General (DG)- The **DG** shall preside at all meetings of the Board. In the event of a vacancy in the position of Chair, the Board shall either at their next meeting or within sixty (60) days of the vacancy, whichever event should come first, fill the vacancy from inside the Board's own number for the unexpired term.

2. The Deputy Director General (DDG) - The **DDG** shall have such powers and perform such duties as the Board or the Chair may assign. In the absence or inability of the Chair to act, the Co-Chair shall perform all the duties and exercise all the powers of the Chair.

3. The Secretary- The Secretary shall keep the minutes of all meetings of the Board, issue proper notices of all Board meetings, have charge of the seal of the Initiative and of such books and papers as the Board may direct, and maintain and authenticate the records that the Initiative is required to maintain at its principal office under applicable law. The Secretary shall have other duties as shall be assigned by the Board from time to time.

4. The Treasurer- The Treasurer shall collect all monies due to SLAM and have custody of the funds of SLAM same in such depositories as may be approved by the Board. At each annual meeting of the Board and at such other times as the Board may require, the Treasurer shall submit to the Board a report of the financial condition of the Initiative. The Treasurer shall perform such other duties as may be assigned by the Board.

5. Project Coordinator/Fundraising and Deputy:
Functions/Roles of project coordinator and deputy as well as project implementation.

Project/Fundraising Coordinator

SLAM will, from time to time, undertake specific projects with the goal of improving the lives and livelihood of the people of Sierra Leone and its members. To that end, SLAM will elect/appoint a Project Coordinator (PC) and a Deputy who will oversee projects to meet the specific identified needs of Sierra Leoneans in consultation with the Executive of SLAM.

The Project Coordinator is responsible for fundraising and overseeing the development and execution of project deliverables, leading project planning, resource mobilization, budgeting, action plans, and monitoring processes. The PC is also responsible for submitting activity reports within the agreed time, meeting minutes, and financial reports on a regular basis to the donor, the Executive, and the membership.

Project Managers (PM) can be appointed for specific projects as necessary, and the PC has the responsibility to assist the project manager's teams with the coordination of resources, equipment, meetings, and information. They organize projects with the goal of getting them completed on time and within budget.

Key role/description

The project coordinator will coordinate the schedule, budget, issues, and risks of the project. It's their job to ensure the project management framework is well-organized and that it runs smoothly. This can include communicating with various departments in the organization to ensure everyone is on the same page.

Responsibilities

a. Plan, budget, oversee and document all aspects of a specific project

b. Organizing, attending, and participating in stakeholder meetings.

c. Ensure stakeholder views/expectations are managed to achieve the best solution.

d. Work closely with the Executive to ensure that the scope and direction of each project are on schedule.

e. Providing administrative support to PM as needed.

f. Assess project risks and issues and provide solutions where applicable.

g. Coordinate and complete *projects* on time within budget and within scope

h. Monitor and summarize the progress of *projects* and submit final reports to project managers and stakeholders.

i. Assisting with resource scheduling so team members have the resources they need to complete their tasks.

j. Scheduling stakeholders' meetings and facilitating communication between the project managers and stakeholders throughout the project life.

k. Managing project management documents such as the project plan, budget, and schedule or scope statement, as directed by the project manager.

l. Executing a variety of project management administrative tasks such as billing and bookkeeping.

m. Support team members when implementing risk management strategies.

Project Coordinator Requirements:

a. Minimum Bachelor's degree in business or related field of study.

b. Three years of experience in a related field with a proven record of delivering on projects

c. Exceptional verbal, written, and presentation skills.

d. Good people skills capable of maintaining strong relationships.

e. Strong organizational and multi-tasking skills.

f. Excellent analytical and problem-solving abilities.

g. Strong leadership skills

h. Ability to work effectively both independently and as part of a team.

i. Experience using computers for a variety of tasks.

j. Competency in Microsoft applications, including Word, Excel, and Outlook.

k. Knowledge of file management, transcription, and other administrative procedures.

l. Ability to work on tight deadlines.

5. (1) Deputy Project Coordinator (DPC) and Functions

The DPC is the principal assistant to the PC. S/he shall assume the roles and responsibilities of the substantive holder during his/her absence. Such an assumption of duty is of right and immediate.

6. Auditors - Functions/Roles

The auditors shall review, audit, and provide a professional and independent true and fair view of the organization's financial statements/accounts.

Amongst all other roles and responsibilities, the auditors shall perform the following functions:

a. Determine the accuracy of the financial statements prepared by members or committees

b. Analyse and advise going concerns of the organization

c. Detect and prevent any fraud and discrepancies in the organization's accounting systems, as well as investigate instances of possible fraud (even those considered immaterial)

d. Ensure that the organization complies with its legal obligations

e. Report matters of all material significance to the chairman or advisory board.

f. Provide recommendations to improve weak internal controls.

g. Perform reconciliations of financial and operating information.

h. Monitor compliance with industry standards, laws, and guidelines.

The auditors' objectives are to obtain reasonable assurance about whether the financial statements as a whole are free from material misstatement, whether due to fraud or error and to issue an auditor's report that includes the auditor's opinion. Reasonable assurance is a high level of assurance but is not a guarantee that an audit conducted in accordance with International Standards on Auditing (UK) (ISAs (UK)) will always detect a material misstatement when it exists. Misstatements can arise from fraud or error and are considered material if, individually or in the aggregate, they could reasonably be expected to influence the economic decisions of users taken based on these financial statements.

It is the responsibility of the management committee to provide the financial statements. The role of the auditors or reviewers is to give a professional and independent **advice** on these financial statements. The review or audit of an association's financial report can ensure greater accountability to the members and provide an assurance that all funds received by the organization have been correctly accounted for.

The committee should not rely on the auditor to find all errors in the statements and identify any fraud. It remains the committee's responsibility to pay close attention to the association's financial statements at all times.

The auditor's task is to provide a professional opinion on the state of the financial affairs of the association. Auditors have a legal responsibility for their opinion and can be held liable for negligence if the audit is not completed according to professional standards or for damage to the association as a result of negligence.

What if the audit report is unfavourable?

There is always the possibility **that** an auditor may present a critical report identifying areas that the association needs to address. To ignore an auditor's report is likely to place the association at risk.

If the association is unsure concerning the auditor's statement, **SLAM** should seek clarification as irregularities in the financial statements could occur for a number of reasons, including:

a. a lack of understanding in preparing financial statements;
b. a lack of understanding in assessing financial statements;
c. poor controls over money in and out, or
d. Dishonesty.

Suppose problems suggesting dishonesty are found in the financial records, the association should obtain prompt legal advice and attend to any immediate matters such as freezing accounts, securing assets, investigating, and contacting the police and/or the insurer.

7. Public Relationship Officer (PRO)

The PRO shall be responsible for the public good of the ASSOCIATION. His/her duties and obligations may include the following:

    a. Identify platforms for marketing 8.1(c) Promoting and boosting the modus operandi of the association 8.1(d) Keeping the banner of the association high

    b. He/she shall be liaising with and providing answers to public inquiries from media, individuals, and other organisations, often via telephone, email, and other digital platforms.

    c. He/she will maintain and update information on the organisation's website, and support in the organisation of events, including holding press conferences, etc.

    d. identify platforms for marketing

    e. Promoting and boosting the modus operandi of the association

    f. Keeping the banner of the association high

7. (1) Deputy Public Relationship Officer

The Deputy Public Relationship Officer is the principal assistant to the PRO.

S/he shall assume the roles and responsibilities of the substantive holder during his/her absence.

Such an assumption of duty is of right and immediate.

8. Organising Secretary

S/he shall be responsible for organising charitable activities. He or she shall oversee all social/charitable activities of the association. Whenever the association shall engage in charitable functions, s/he shall head the committee for various duties assigned to members of such committee. All such events as talks, seminars, training workshops, fund-raising, project proposals meetings, visitation of sites, places, and locations where planned charitable work is to be carried, out and certain social functions associated with the SLAM.

8. (1) Deputy Organising Secretary

The Deputy Organising Secretary is the principal assistant to the Organising Secretary.

S/he shall assume the roles and responsibilities of the substantive holder during their absence. Such an assumption of duty is of right and immediate.

9. Welfare/Advisory Council Committee (Advisers)

The key responsibility of the SLAM Welfare/Advisers is to help the association meet its stated goals and objectives, and such shall be full-time members of the association. SLAM advisors shall if deemed necessary:

a. Seek the welfare of SLAM members as shall be prescribed and directed by the Executive/Directors (i.e., those who may resort to marriage and/or experience family bereavement). The eligibility criteria and amount of financial help/support shall be determined by the Board of Directors/Executive members

b. Offer advice on agenda items

c. Advise on the planning of activities

d. Help resolve problems in the association or mediate in personality and other conflicts

e. Contribute towards team building in the SLAM membership

f. Assist both old and new leadership in times of transition/change of administration to ensure continuity

g. Any other advisory roles relevant to the achievement of SLAM objectives

10. Such other Director(s) as the Board/Executive may from time to time select by a majority vote of all Directors/Executive in office; provided, however, that with respect to those members serving on the Executive Committee, such members' respective memberships on the Executive Committee shall be for no more than a total of five years. Except that such limitation shall not apply to any of such members who during such memberships on the Executive Committee are elected to any of the Board officer positions identified above, in which case such memberships on the Executive Committee, as applicable, shall be vacated and filled in accordance therewith; and

11. Director General (DG)- In addition to the members of the Executive Committee identified in the foregoing paragraphs, the (DG) shall be the head of the Executive Committee. He/she shall serve as the director of SLAM staff, including supervision, control, hiring, and termination of such staff, for which service he or she may receive reasonable compensation and benefits as determined by the Board in accordance with applicable law, rules, and regulations. He/she may co-sign with the **Deputy Director General**, Secretary, Treasurer, or any other officer of SLAM authorised by the Board, deed, mortgages, bonds, contracts, or other instruments which the Board has authorised to be executed, except in an instance for which the execution thereof shall be expressly delegated by statute, by laws, or the Board to some other officer or agent of the Initiative. The **Director General** may, for the purpose of **Fundraising, Welfare/Advisory,** and other pertinent reasons, establish relevant country branches/offices with the approval of the board.

ARTICLE 17: ADDITIONAL COMMITTEES

1. The Board may delegate to any additional Board committee, including, but not limited to, a Governance Committee, a Budget and Finance Committee, an Audit and Compliance Committee, a Human Resources and Administration Committee, a Program and Business Development Committee, and a Public Affairs Committee. Any of the authority of the Board that it may lawfully delegate, other than the authority to take any action. The creation of additional Board committees beyond those expressly identified in the previous sentence shall require the vote of a majority of all Directors in office.

Building A Nation

2. Additional Board committees shall consist of only Directors and no fewer than three (3), except for the Governance Committee, which shall consist of no fewer than five (5). Appointments of Directors to serve as members of additional Board committees shall require the vote of a majority of all Directors in office.

3. The **Director General** and the **Deputy Director General** shall, in their sole joint discretion, appoint a chair and co-chair (where applicable) for each additional Board committee to chair meetings of such committees and to report to the full Board/Executive on such meetings. All committee members shall be eligible to serve as chair or co-chair.

4. Each such additional Board/Executive committee shall serve at the pleasure of the Board/ Executive and shall be subject to the control and direction of the Board, provided, however, that any third party shall not be adversely affected by relying upon any act by any such committee within the authority delegated to it.

5. At any meeting of an additional Board/Executive committee, the presence of a minimum of the following number of all Directors/Executives serving as members on such committee shall be required to establish a quorum for the transaction of business. The vote of a majority of the members of an additional Board committee present at a meeting at which a quorum is established shall constitute the action of such additional Board/Executive committee.

CHAPTER III: MEMBERSHIP AND ELIGIBILITY, SUSPENSION AND TERMINATION

ARTICLE 18: MEMBERSHIP

1. For the purpose of fund-raising, the Board shall establish membership branches/chapters in relevant countries for interested adults.

2. Membership shall be open and voluntary to every descendant and friend of the Republic of Sierra Leone who wishes to participate in building and sustaining the general welfare and good of the district.

3. Members who deliberately misrepresent SLAM Global through bad conduct, fraud, or the pursuit of personal and financial gain shall be expelled following a two-thirds majority vote of fully registered members.

4. Members whose activities are sine qua non to the occurrence of financial mischief shall be prosecuted for criminal-related offences.

5. By an ethical, moral, and fiscal commitment to the smooth running of the Advocacy and shall therefore be considered based on this commitment by each who wishes to be considered a member of the initiative. A code of Conduct and Ethics manual shall be developed to prescribe the ethical and moral commitment of each and the consequences of breach of any of the prescribed Codes of Conduct and Ethics.

6. Members who embark on verbal and physical aggression against other members, officials, and anybody else with respect to issues concerning SLAM Global shall have their membership reviewed by the Advisory Council. And any decisions adopted shall be binding.

7. Membership is on a social equity basis to ensure mutual respect and pluralism.

ARTICLE 19: FEES AND CONTRIBUTIONS

1. The Board/Executive, in consultation with Global country members, shall determine a one-time registration fee.

2. Membership shall be attained by a registration fee of ($60) where applicable) or its equivalent to be followed by a monthly contribution of ($10). Members who fail to make the monthly contribution for four consecutive months may be suspended, temporarily prevented from voting, and will be required to provide evidence of mitigating or exceptional circumstances in order to be reinstated.

3. Membership fees, dues, and all costs shall be determined from time to time by the Board.

ARTICLE 20: ELIGIBILITY

Eligibility for executive positions in members' chapters/branches

1. Be a registered member of the SLAM chapter.

2. Abide by this constitution and chapter bylaws; respect its letters, notices, Authorities, and offices established under this constitution or any other law passed by the Advocacy; and

4. Ensure that he/she attends general meetings

ARTICLE 21: SUSPENSION AND TERMINATION

1. Subject to the provisions outlined in this constitution, the Code of Conduct and Ethics manual, and any other rules of the Initiative and subject to an investigation by a constituted ad hoc Committee acting on that behalf of and on its recommendations, membership shall be suspended.

2. If any member is considered to be guilty of failing in his or her duty as a member after several warnings and or a suspension, the membership of a member can be terminated after due process by the ad hoc committee subject to an appeal to the advisory body; and

3. Procedures for Reinstatement of Terminated membership shall be prescribed clearly in the Initiative's Code of Conduct and ethics.

CHAPTER IV: INDEMNITY

ARTICLE 22

1. The Initiative shall indemnify and keep indemnified every officer, member, volunteer, and employee of SLAM from and against all claims, demands, actions, and proceedings (and all costs and expenses in connection therewith or arising therefrom) made or brought against the Initiative in connection with its activities, the actions of its officers, members, volunteers or employees, or in connection with its property and equipment. This indemnity shall not extend to liabilities arising from wilful and individual fraud, wrongdoing, or wrongful omission on the part of the officer, member, volunteer, or employee sought to be made liable; and

2. The Treasurer shall affect a policy of insurance with respect to this indemnity.

CHAPTER V: GENERAL PROVISIONS

ARTICLE 23: NOTICES

Whenever, under applicable law, notice is required to be given to any person, such requirement shall not be construed to mean personal notice to such person, and notice shall be sufficient if it is given in writing by:

1. Mail or an internationally recognized express carrier, addressed to such person at their address as it appears on the records of the SLAM, with postage and delivery charges thereon prepaid. Such notice shall be deemed to be given at the time when the same shall be deposited in the mail or with the carrier;

and

2. Notice may also be given by facsimile or email, addressed to such person to the number or email address appearing on the records of the Initiative; provided, however, that a notice by such additional methods shall not be deemed to have been given to such person unless and until such person receives such notice, but regardless of whether such person actually reads such notice.

ARTICLE 24: WAIVER OF NOTICE

1. Without limitation to Article 24, Paragraph 2, whenever notices are required under Paragraph 1 of Article 24, a waiver thereof in writing signed by the person or persons entitled to such notice, whether before or after the time stated therein, shall be equivalent thereto.

ARTICLE 25: FISCAL YEAR

The fiscal year of SLAM shall be as fixed by the action of the Board.

ARTICLE 26: LEGAL COUNSEL

Legal counsel to SLAM may be appointed by action of the Board, and such legal counsel shall render to the Board and its designees such legal advice as may be necessary and convenient in the operations of the Initiative.

CHAPTER VI: AMENDMENTS

ARTICLE 27: AMENDMENTS

This Constitution and any other law subject thereto may be amended or repealed at any time by action of the Board, provided that no less than ten (10) days notice of intent to so amend or repeal has been given to each Director.

CHAPTER VII: LEGAL STATUS

ARTICLE 28: LEGAL STATUS

1. Subject to the provisions of the Law of the country of registration and under any applicable law, this constitution shall bind the members of the Initiative, inclusive of those in branches/chapters.

2. The laws of the Initiative include this constitution, bylaws, policies, codes of conduct, ethics manual, rules, regulations, and any other law that the Board may so institute to regulate the conduct of its members.

3. The Initiative is a juristic person and is the bearer of its own rights and responsibilities. It can sue and can be sued.

4. The Initiative is registered as a limited by guarantee corporate entity; and

5. The constitution shall have no legal force and effect unless such constitution and amendments thereto are approved by the respective bodies.

CHAPTER VIII: REGULAR REVIEWS AND UPDATES

ARTICLE 29: REGULAR REVIEW CYCLE

1. **Review Schedule**: The Constitution shall be reviewed biennially (every two years) to ensure its relevance and effectiveness in guiding SLAM. The first review shall occur two years from the date of adoption of this Constitution.

2. **Review Committee**: The Board shall establish a Review Committee at least six months before the scheduled review date. This committee shall comprise a diverse group of Directors and members from various chapters to provide comprehensive and inclusive feedback.

ARTICLE 30: REVIEW PROCESS

1. **Stakeholder Engagement**: The Review Committee shall engage with stakeholders, including members, chapter leaders, beneficiaries, and external partners, to gather feedback on the Constitution. This engagement shall include surveys, focus groups, and public forums.

2. **Assessment Criteria**: The Review Committee shall assess the Constitution based on its alignment with SLAM's mission, vision, values, and the evolving needs of the organization and its stakeholders. Consideration shall also be given to changes in the legal, social, and political environments in which SLAM operates.

3. **Recommendations**: The Review Committee shall compile a report detailing their findings and recommendations for amendments or updates to the Constitution.

This report shall be presented to the Board for discussion and approval.

ARTICLE 31: AMENDMENT PROCEDURE

1. **Proposal of Amendments**: Amendments to the Constitution may be proposed by the Review Committee or any Director. Proposals must be submitted in writing and include a rationale for the changes.

2. **Board Approval**: Proposed amendments shall be reviewed and approved by a two-thirds majority vote of the Board. Approved amendments shall be documented and communicated to all members and stakeholders.

3. **Member Ratification**: Significant amendments, as determined by the Board, shall be presented to the general membership for ratification. A special meeting may be called for this purpose, and amendments shall be ratified by a simple majority vote of the members present.

ARTICLE 32: COMMUNICATION OF UPDATES

1. **Notification**: All members shall be notified of any amendments or updates to the Constitution within thirty (30) days of their approval. Notification methods shall include email, official SLAM website announcements, and chapter meetings.

2. **Accessibility**: The updated Constitution shall be made accessible to all members through the SLAM website, and copies will be distributed to chapter leaders.

A summary of significant changes shall be provided to ensure members are aware of how the updates may affect their roles and responsibilities.

ARTICLE 33: CONTINUOUS IMPROVEMENT

1. **Ongoing Feedback Mechanism**: SLAM shall establish an ongoing feedback mechanism, such as an online platform, where members can continuously provide suggestions for improvement. This platform shall be monitored regularly by the Review Committee or a designated body.

2. **Interim Reviews**: In addition to the biennial reviews, the Board may initiate interim reviews in response to significant organizational changes or external events that necessitate immediate updates to the Constitution.

34. ARTICLE 18 - ARBITRATION

1. In the case of serious disagreement over the interpretation of this constitution between the executive on the one hand and the rest of the SLAM on the other hand and between various factions in SLAM, the matter will be referred to the association's legal adviser, whose interpretation shall be final.

35. ARTICLE 19- BINDING AGREEMENT

1. If any part of these promises is void for any reason, the undersigned accepts that it may be severed without affecting the validity or enforceability of the balance of the promises.

2. Any member of SLAM, on becoming an executive committee member is hereby empowered to actively adhere to all SLAM contributions.

3. This agreement shall be binding upon and inure to the benefit of SLAM Global parties, their successors, assignee(s), and personal representatives.

36. ARTICLE 20- IN WITNESS WHEREOF, each party to this agreement has caused it to be executed in place and on the date indicated below.

CHAPTER IX: GLOSSARY OF TERMS

To ensure clarity and a common understanding of key terms and roles within SLAM, the following glossary provides definitions for terms used throughout this Constitution.

ARTICLE 34: DEFINITIONS

1. **Advocacy**: Activities are undertaken to influence public policy and resource allocation decisions within political, economic, and social systems and institutions.

2. **Board of Directors (Board)**: The governing body of SLAM responsible for overseeing the organization's activities, setting strategic direction, and ensuring the organization's mission is fulfilled.

3. **Non-Political**: Not affiliated with, influenced by, or supporting any political party or candidate. SLAM operates independently of any political agenda to ensure unbiased advocacy and service to the community.

4. **Secular**: Not connected with religious or spiritual matters. SLAM is inclusive and neutral with respect to religious beliefs, ensuring that its activities and policies do not promote or endorse any particular religion.

5. **Bylaws**: The set of rules or regulations established by SLAM that govern its operations and the behavior of its members.

6. **Chair**: The leader of the Board who presides over Board meetings and represents the Board in interactions with external stakeholders.

7. **Chapter**: A local or regional branch of SLAM established to carry out the organization's mission and activities within a specific geographical area.

8. **Conflict of Interest**: A situation where a member's personal or professional interests may interfere with their ability to act in the best interests of SLAM.

9. **Diaspora**: The community of Sierra Leoneans living outside their homeland who maintain connections and contribute to the development of Sierra Leone.

10. **Director**: A member of the Board who is elected to provide governance and oversight for SLAM.

11. **Ethics**: The moral principles that govern a person's behavior or the conducting of an activity, particularly within SLAM.

12. **Executive Committee**: A subset of the Board with specific delegated authority to manage SLAM's affairs between Board meetings.

13. **Ex officio**: A member of a body (such as the Board or a committee) who is part of it by virtue of holding another office.

14. **Fiscal Year**: The 12-month period used for accounting purposes and preparing financial statements, as determined by the Board.

15. **Governance Committee**: A Board committee responsible for overseeing the organization's governance practices and ensuring compliance with its Constitution and Bylaws.

16. **Indemnity**: Protection against financial loss or liability provided to SLAM officers, members, volunteers, and employees in connection with their duties.

17. **Initiative**: Refers to SLAM as an organized effort to achieve specific goals, particularly in advocacy, development, and community engagement.

18. **Member**: An individual who has registered with SLAM and contributes to its activities and mission according to the established guidelines.

19. **Mission**: The core purpose of SLAM, describing what the organization seeks to achieve and how it aims to impact the community.

20. **Officer**: A person holding a position of authority within SLAM, such as Chair, Co-Chair, Vice Chair, Secretary, or Treasurer.

21. **Quorum**: The minimum number of members or Directors required to be present at a meeting to conduct official business.

22. **Secretary**: The officer responsible for maintaining the official records of SLAM, including minutes of meetings and official correspondence.

23. **SLAM**: The Sierra Leone Advocacy Movement, an organization dedicated to advocating for the rights and welfare of Sierra Leoneans and promoting democratic principles.

24. **Stakeholder**: Any individual, group, or organization that has an interest or concern in SLAM and its activities.

25. **Treasurer**: The officer responsible for managing SLAM's finances, including collecting dues, maintaining financial records, and providing financial reports.

26. **Vision**: The long-term aspiration of SLAM, describing the desired future state and impact of the organization.

27. **Vice Chair**: An officer who assists the Chair and may perform the Chair's duties in their absence or inability to act.

28. **Voting**: The act of expressing a choice or decision in a formal setting, such as a Board or member meeting, to make official decisions.

29. **Volunteer**: An individual who contributes their time and skills to support SLAM's activities without receiving monetary compensation.

30. **Affiliate Membership**: A type of membership in SLAM that allows individuals or organizations to support and participate in SLAM's activities without the full responsibilities and rights of regular members. Affiliate members may contribute to specific projects, attend events, and offer expertise but do not have voting rights or eligibility for leadership positions within the organization.

Summary and Call to Action

Expanded Conclusion: In "**Building a Nation: Good Governance, Trust Building and Democratization in Sierra Leone**", we have journeyed through the intricate landscape of Sierra Leone's political evolution, governance challenges, and the pivotal role of democratic principles in fostering a just and prosperous society. This book has delved into various governance principles, providing a comprehensive understanding of their importance and application in Sierra Leone's context.

Key Takeaways:

1. **Historical Context:**
    - Sierra Leone's history of governance is marked by a transition from colonial rule to independence, followed by periods of political instability and conflict.
    - The civil war (1991-2002) highlighted the severe consequences of governance failures, emphasizing the need for robust democratic institutions and processes.

2. **Principles of Good Governance:**
    - **Rule of Law:** Ensuring that laws are applied consistently and fairly to all citizens.
    - **Transparency:** Government actions and decisions must be open to public scrutiny to foster trust and accountability.
    - **Accountability:** Leaders must be accountable to the people, with mechanisms in place to hold them responsible for their actions.
    - **Participation:** Inclusive participation of all citizens in the political process is crucial for a healthy democracy.
    - **Equity and Inclusiveness:** Ensuring that all groups, especially marginalized ones, have opportunities to participate and benefit from development.

3. **Current Challenges and Opportunities:**

- Despite progress in rebuilding after the civil war, Sierra Leone faces ongoing issues such as corruption, economic underdevelopment, and social inequalities.
- The Sierra Leonean diaspora, through organizations like SLAM, plays a vital role in supporting national development and promoting democratic values.

4. **The Role of SLAM**:
    - SLAM's vision of a vibrant, democratic Sierra Leone aligns with the national coat of arms' values of unity, freedom, and justice.
    - The organization's mission includes mobilizing diaspora resources, advocating for social justice, promoting gender parity, and fostering democratic engagement.

Reflection and Engagement Questions

This chapter focuses on the role of SLAM in promoting good governance and democratic principles. It highlights SLAM's efforts in advocating for information laws, developing information platforms, training journalists and civic groups, conducting public awareness campaigns, and partnering with international transparency organizations. By exploring these initiatives, the chapter aims to demonstrate how strategic advocacy and grassroots mobilization can drive significant changes in governance and transparency.

The chapter also emphasizes the importance of sustained efforts and collaboration with various stakeholders to achieve long-lasting impact.

1. **SLAM's Contributions:**
   - How has SLAM's advocacy for information laws contributed to increased transparency and accountability in Sierra Leone?

   - _____

2. **Development of Information Platforms:**
   - What are the benefits of developing information platforms for journalists and civic groups in promoting good governance?

   - _____

3. **Training and Capacity Building:**
   - Reflect on the importance of training journalists and civic groups. How do these efforts enhance the effectiveness of SLAM's initiatives?

   - _____

4. **Public Awareness Campaigns:**

    o   How can public awareness campaigns be used to educate citizens about their rights and responsibilities in governance? Provide examples of successful campaigns.

    o   _____
        _____

5. **International Partnerships:**

    o   In what ways can partnerships with international transparency organizations strengthen SLAM's advocacy efforts and overall impact on Sierra Leone's governance?

    o   _____
        _____

These questions are designed to help readers engage deeply with the chapter's content, promoting critical thinking and encouraging them to consider practical solutions and personal contributions to the advocacy movement in Sierra Leone.

CHAPTER 11:
OVERALL CONCLUSION AND CALL TO ACTION

As we conclude this exploration of governance and democracy in Sierra Leone, it is imperative to recognize that the journey towards a fully democratic and prosperous society is ongoing. The principles discussed in this book are not merely theoretical constructs but actionable guidelines that can transform Sierra Leone's political landscape when effectively implemented.

Call to Action for Readers:

1. **Get Involved in Governance Issues**:
    - Stay informed about local and national political developments. Knowledge is a powerful tool for advocating change.
    - Participate in public consultations, town hall meetings, and other forums where governance issues are discussed.

2. **Support Advocacy Movements**:
    - Join or support organizations like SLAM that are committed to promoting good governance, social justice, and democratic principles.

- Volunteer your time, skills, or resources to advocacy groups working towards meaningful change in Sierra Leone.

3. **Engage in Community Dialogues**:
    - Foster dialogue within your community about the importance of democratic values and good governance.
    - Encourage others to participate in civic activities and emphasize the importance of collective action in holding leaders accountable.

4. **Promote Transparency and Accountability**:
    - Demand transparency from government officials and institutions. Use available tools and platforms to monitor government actions and expenditures.
    - Support initiatives that aim to enhance accountability, such as investigative journalism and watchdog organizations.

5. **Champion Equity and Inclusiveness**:
    - Advocate for policies and practices that promote equality and inclusiveness in all sectors of society.
    - Support initiatives aimed at empowering marginalized groups, ensuring their voices are heard in the political process.

Final Thoughts

The future of Sierra Leone depends on the active participation and commitment of its citizens, both at home and in the diaspora. By embracing the principles of good governance and democratic engagement, we can collectively work towards building a nation that truly embodies the values of unity, freedom, and justice.

In the words of Nelson Mandela, "Education is the most powerful weapon which you can use to change the world." Let us educate ourselves and others, engage in meaningful action, and strive to create a Sierra Leone where every citizen can thrive and contribute to the nation's progress. Your involvement can make a significant difference in shaping the future of Sierra Leone.

Reflection and Engagement Questions

As we conclude this exploration of governance and democracy in Sierra Leone, it is imperative to recognize that the journey towards a fully democratic and prosperous society is ongoing. The principles discussed in this book are not merely theoretical constructs but actionable guidelines that can transform Sierra Leone's political landscape when effectively implemented. This chapter serves as a call to action for readers to engage actively in governance issues, support advocacy movements, participate in community dialogues, promote transparency and accountability, and champion equity and inclusiveness. By taking these steps, we can collectively work towards building a nation that embodies the values of unity, freedom, and justice.

1. **Personal Involvement:**
   - How can you personally get involved in addressing governance issues in Sierra Leone? What specific actions can you take to make a difference?

   - _____
     _____

2. **Supporting Advocacy Movements:**
   - Reflect on the role of advocacy movements like SLAM. How can you support these movements to promote good governance and democratic principles?

   - _____
     _____

3. **Community Engagement:**
   - Why is it important to engage in community dialogues about democratic values and good governance? What strategies can you use to foster these discussions in your community?

   - _____
     _____

4. **Promoting Transparency and Accountability:**
    - What practical steps can you take to promote transparency and accountability within your local government or organization?

    - _____
      _____

5. **Championing Equity and Inclusiveness:**
    - How can you advocate for policies and practices that promote equity and inclusiveness in Sierra Leone? Provide examples of initiatives that could help achieve these goals.

    - _____
      _____

These questions are designed to help readers reflect on the chapter's content , encouraging them to think critically about their role in promoting democratic governance and consider practical actions they can take to contribute to Sierra Leone's development.

ACKNOWLEDGMENTS

The creation of this book, "Building a Nation - Good Governance, Trust Building and Democratization in Sierra Leone," has been a collaborative effort involving many individuals whose contributions have been invaluable.

First and foremost, we would like to express our deepest gratitude to the Sierra Leonean people, both at home and in the diaspora, whose resilience, courage, and unwavering commitment to democracy continue to inspire and drive the quest for good governance.

We extend our heartfelt thanks to the members of the Sierra Leone Advocacy Movement (SLAM). Your dedication to promoting social justice, democratic principles, and national development has been instrumental in shaping the ideas and insights presented in this book. Special thanks to Dr. Alfred Veenod Fullah, whose guidance, feedback, and unwavering support were crucial throughout this journey.

To our families and friends, your constant encouragement and patience have been our pillar of strength. Thank you for believing in this project and supporting us every step of the way.

We are immensely grateful to the numerous experts, scholars, and practitioners whose research and work have enriched this book. Your contributions to the fields of governance, democracy, and development provided the foundational knowledge and perspectives that underpin this manuscript.

A special note of appreciation goes to my editor and proofreader, Dr. Francis Mbunya, for his meticulous attention to detail and insightful suggestions. Your efforts have significantly enhanced the quality and readability of this book.

We would also like to acknowledge the assistance provided by ChatGPT; an AI language model developed by OpenAI. The insights and suggestions generated by ChatGPT have been valuable in refining the content and structure of this book, ensuring clarity and coherence in the presentation of complex ideas.

Finally, we would like to acknowledge the countless individuals who have participated in governance and democratic processes in Sierra Leone, often at great personal risk. Your courage and dedication are the true testament to the power of democracy and the promise of a brighter future for Sierra Leone.

To all who have contributed directly or indirectly to the realization of this book, your support and encouragement have made this journey possible. This book is dedicated to you and the hope that together, we can continue to build a Sierra Leone that truly embodies the values of unity, freedom, and justice.

Thank you.

ABOUT THE AUTHORS

Author's Bio: Dr. Robert B. Kargbo is a passionate advocate for good governance and democratic principles in Sierra Leone, bringing extensive experience in leadership and civic engagement. As the lead author of this book, Dr. Kargbo aims to inspire collective action towards creating a transparent, accountable, and inclusive society in Sierra Leone, grounded in the values of democracy and ethical governance.

In addition to his work in governance, Dr. Kargbo is a leading figure in synthetic organic chemistry, having made significant contributions to the development of bioactive pharmaceutical and psychedelic compounds, including psilocybin and 5-MeO-DMT, for clinical use. His work at the Usona Institute has been pivotal in optimizing chemical synthesis processes, enabling large-scale manufacturing for clinical trials and advancing the therapeutic potential of these compounds.

Dr. Kargbo's scientific career is marked by extensive collaborations with industry and academia, including partnerships with the NIH National Institute of Neurological Disorders and Stroke under the Medicinal Chemistry for Neurotherapeutics Program and Orphan Drug Program. He served as lead Scientist in the Eli Lilly/AMRI In Sourcing Program. He serves on the editorial board of the ACS Medicinal Chemistry Letters and regularly contributes to the Organic Process Research and Development journal.

The lead author of over 250 peer-reviewed publications, several books, and patents to his name, Dr. Kargbo is a distinguished scientist recognized with awards such as the Pfizer Global Research and Development Research Fellowship and the McNair Scholar Award. He earned his PhD in Synthetic Organic Chemistry from North Dakota University and has lectured at the College of Medicine and Allied Health Sciences in Freetown, Sierra Leone.

Dr. Kargbo's latest book, *The Incredible Power of Becoming Your True Self*, delves into personal success stories and the proven strategies he and his co-authors have used to achieve and sustain business success. This work goes beyond anecdotes, offering a practical, step-by-step guide designed to help readers overcome fear and discouragement—two significant hurdles on the path to success. The book reflects Dr. Kargbo's commitment to personal development and his belief in the transformative power of authenticity and resilience.

Beyond his professional achievements, Dr. Kargbo is passionate about fostering a culture of lifelong learning and community service.

He enjoys table tennis, biking, reading, mindfulness and writing, activities that reflect his commitment to personal well-being and intellectual growth.

Dr. Kargbo's dual expertise in scientific research and governance advocacy, combined with his recent literary work, positions him as a multifaceted leader. His contributions to this book underscore his vision for a better Sierra Leone, where democratic principles guide national progress and scientific innovation contributes to the well-being of society.

Author's Bio: Dr. Alfred Amadu Veenod Fullah is a seasoned advocate for human rights and good governance, with a distinguished career that spans education, labor advocacy, and community development both in Sierra Leone and the United Kingdom. Currently serving as the Director-General of the Sierra Leone Advocacy Movement (SLAM-Global), Dr. Fullah has been a pivotal figure in promoting democratic principles and fostering a culture of accountability and transparency in governance.

Dr. Fullah's journey in advocacy began in Sierra Leone, where he worked as a teacher and senior trade unionist, eventually being elected as Vice President of the Sierra Leone Teachers Union (SLTU) for over a decade. His leadership in the union underscored his deep commitment to improving the welfare of educators and advocating for fair labor practices, which laid the groundwork for his later work in peacebuilding and community education.

Following the end of Sierra Leone's civil war, Dr. Fullah took on the role of Project Coordinator for the American Centre for International Labour Solidarity (ACILS) in collaboration with the Sierra Leone Labour Congress (SLLC). In this capacity, he played a crucial role in initiatives aimed at rebuilding the nation through the promotion of peace, good governance, and civic education. His efforts have left a lasting impact on the country's post-war recovery and the strengthening of its democratic institutions.

In addition to his work in Sierra Leone, Dr. Fullah has held various leadership roles in the UK, including Chairman of Like Minds for Makeni - Global, Legal Adviser, and briefly Chairman of the APC UK/I Branch. He has also been actively involved in local politics as the Chair of the Branch Labour Party in Witney, West Oxfordshire, where he successfully campaigned for improved community safety measures.

Dr. Fullah's academic credentials are as impressive as his professional achievements. He holds a master's degree in social policy, an LLM in International Human Rights Law, and a PhD in International Criminal Law, all from Oxford Oxon, UK. His academic work complements his advocacy, providing a robust theoretical foundation for his practical efforts in promoting justice and human rights.

As an author, Dr. Fullah brings a wealth of knowledge and experience to this book. His contributions reflect his unwavering dedication to the principles of democracy, human rights, and the rule of law. Through his writing, he shares valuable insights into the challenges and opportunities facing Sierra Leone as it strives to build a more just and democratic society.

Author's Bio: Mohamed Boye Jallo Jamboria is a distinguished social scientist, professional educator, and dedicated advocate for social justice, with a long history of service to the government of Sierra Leone from 1979 to 1998. His extensive career in public service and education reflects his deep commitment to the advancement of democratic values and the empowerment of communities through education and equitable governance.

Jallo Jamboria's academic journey began at the University of Sierra Leone, where he earned a Bachelor of Arts in Education in 1979. He further enhanced his expertise with two postgraduate diplomas in Human Resource Management and Social Justice, obtained from prestigious institutions supported by the International Labour Organisation in Harare, Zimbabwe, and Buea, Cameroon.

These qualifications underpin his robust understanding of the intersection between education, human rights, and social equity.

His leadership within the Sierra Leone Teachers Union, where he served as Vice President from 1990 to 1992, highlights his lifelong commitment to advocating for the rights of educators and the broader workforce. His work has consistently focused on creating a just and inclusive society, where every individual can thrive.

As an author, Mohamed Boye Jallo Jamboria brings a wealth of knowledge and experience to discussions on good governance and democratic principles. His contributions to this book are rooted in decades of practical experience and a deep understanding of the socio-political challenges that Sierra Leone faces. Through his writing, he offers valuable insights into the mechanisms of governance and the vital role of education in fostering a democratic society.

Author's Bio: Mrs. Hassanatu K. Turay is a multifaceted leader deeply committed to driving positive change in her community and beyond. As a mother, entrepreneur, philanthropist, and political activist, she embodies a rare combination of compassion, innovation, and dedication.

Holding a Bachelor of Science in Nursing, Mrs. Turay merges her passion for healthcare with her entrepreneurial ventures, striving to empower others through innovative business solutions. Her philanthropic efforts are guided by a strong spiritual foundation, as she believes in the power of compassion and kindness to uplift those in need.

Her entrepreneurial drive fuels her pursuit of innovative solutions that enhance the well-being of those around her, while her philanthropic endeavors underscore her steadfast dedication to social justice and the empowerment of her community. In the political arena, Mrs. Turay is a passionate advocate for policies that advance health, education, and equality, using her platform to champion the rights of the underserved and marginalized. Faith plays a central role in her life, guiding her actions and imbuing her work with a profound sense of purpose.

Mrs. Turay is also a fervent advocate for social justice, tirelessly engaging in political and community initiatives to create a just and equitable world. Her work is rooted in a deep commitment to unity, peace, and justice, reflecting her desire to inspire, uplift, and serve others. Mrs. Hassanatu K. Turay has become a beacon of hope and a force for positive change in Sierra Leone and beyond through her diverse endeavors.

Author's Bio: Alieu M. Bah is a dedicated advocate for democratic governance and civic responsibility, with roots in Gbendembu, Sierra Leone. His academic journey began at Saint Francis Secondary School in Makeni, where he developed a strong foundation in the sciences, graduating with A-Levels in 1993. Alieu furthered his education by earning a certificate in Business Management from the Anhalt University of Applied Sciences in Koethen, Germany, and is currently pursuing part-time studies in Business Management at the University of East London (UEL).

Residing in London, UK, Alieu balances his professional life in the retail sector with a deep commitment to the principles of good governance and advocacy for his homeland. His extensive experience as a trained warehouse operative has honed his skills in organization and management, which he channels into his civic endeavors.

Alieu's contributions to this book reflect his lifelong dedication to the cause of democracy in Sierra Leone. His writings are informed by both his personal experiences and his academic pursuits, offering a practical perspective on the challenges of implementing democratic reforms in a post-conflict society.

A passionate football enthusiast, music lover, and avid reader, Alieu's approach to governance is rooted in the belief that a well-informed and engaged citizenry is the cornerstone of a thriving democracy.

As an author, Alieu M. Bah brings a unique blend of practical insights and a steadfast commitment to advocating for a Sierra Leone where democratic principles are upheld, and the rights of all citizens are respected and protected.

Author's Bio: Marian Kamara is a distinguished advocate for social justice and democratic governance, with a career dedicated to uplifting the voices of marginalized communities in Sierra Leone and beyond. With an extensive background in policy analysis, civic education, and grassroots mobilization, she has been at the forefront of numerous initiatives aimed at fostering inclusive political participation and ensuring government accountability.

Marian's work reflects a deep commitment to the principles of democracy, human rights, and gender equality, which are central to the vision of a unified and just Sierra Leone.

Her contributions have been instrumental in shaping public discourse around electoral integrity, transparency, and the rule of law, making her a pivotal figure in the country's ongoing journey towards sustainable democratic governance.

As an author, Marian brings a wealth of experience from her engagements with both national and international organizations, where she has consistently championed the need for ethical leadership and citizen empowerment. Her insights in this book draw from years of on-the-ground experience and a profound understanding of Sierra Leone's socio-political landscape, offering readers a nuanced perspective on the challenges and opportunities in building a truly democratic society.

Marian Kamara holds a degree in Political Science and International Relations and has been recognized for her unwavering dedication to advancing democratic principles. Her writing in this volume not only reflects her expertise but also her passion for creating a future where every Sierra Leonean can fully participate in the democratic process, free from fear and with the full protection of their rights.

Author's Bio: Dr. Mohamed Sesay is a medical doctor based in the United Kingdom, whose passion for philanthropy and social justice has driven his commitment to fostering positive change in Sierra Leone, his country of birth. Beyond his medical practice, Dr. Sesay has dedicated significant efforts to supporting and building communities, particularly focusing on initiatives that promote social equity and justice.

His work reflects a deep-seated belief in the power of community-driven development and the importance of creating equitable opportunities for all citizens. Dr. Sesay's philanthropic endeavors have been instrumental in advancing healthcare, education, and social welfare in Sierra Leone, making him a respected figure in both the diaspora and his homeland.

As an author, Dr. Sesay brings a unique perspective to the discourse on good governance and democratic principles. His contributions to this book are informed by his dual experiences in medicine and social advocacy, offering a holistic view of the challenges facing Sierra Leone as it works to strengthen its democratic institutions and improve the well-being of its people.

Married and a proud father of two, Dr. Sesay's dedication to his family is matched by his unwavering commitment to the betterment of Sierra Leone. His writing in this volume is a testament to his belief that sustainable development and social justice are achievable through collective action and the promotion of democratic values.

Author's Bio: Fatima Lydia Sessay, a Sierra Leonean by birth and a British citizen, is an accomplished professional in education and lifelong learning. She is married and the proud mother of two children.

With extensive experience as a qualified Assessor and Quality Assurance Officer, Fatima has been instrumental in delivering training on both accredited and non-accredited programs. She is currently employed as an Apprenticeship Training Coordinator for one of the leading training consortia in the UK.

Fatima holds a postgraduate Master's in Sustainable Development from the University of Middlesex, UK, and a graduate degree in Humanities and Social Sciences from Fourah Bay College, University of Sierra Leone.

Her key interests include the environment, peace, politics, and sustainability. Fatima's dedication to these areas is evident in her professional and personal pursuits, as she continually seeks to contribute to the development and empowerment of her community.

Author's Bio: Mrs. Zainab Melvina Omoyinmi (née Tholley) is a dedicated social worker, educator, and advocate for social justice, with deep roots in Sierra Leone, where she was born and raised. Her journey in education began at the Guadalupe Secondary Vocational School in Lunsar, Port Loko District, where she attained her GCSE O levels. She furthered her education at Milton Margai Teachers College in Freetown, earning a Higher Teachers Certificate (Secondary), which laid the foundation for her career in teaching and social work.

Building A Nation

Driven by a passion for supporting communities and advocating for the vulnerable, Mrs. Omoyinmi pursued higher education in the United Kingdom, obtaining a Bachelor of Arts Honours degree in Social Work from Brunel University. She later achieved a Master's degree in International Social Work & Community Development Studies at East London University. With over 20 years of experience as a qualified and registered social worker, Mrs. Omoyinmi has worked in various senior and management roles within local authorities across the UK, specializing in safeguarding adults.

Currently, she serves as a Consultant Lead Practitioner in Safeguarding Adults and operates as an Independent Consultant Best Interest Assessor under the Deprivation of Liberty Safeguards. Her professional expertise is complemented by her active involvement in the Sierra Leonean community in the UK, where she has held leadership positions, including President of the Guadalupe Old Girls Association in the UK and Chairlady of the Milton Margai Alumni UK.

Mrs. Omoyinmi's commitment to social justice extends beyond her professional life. She is deeply involved in voluntary and charity work, focusing on advocacy, raising awareness on well-being and social issues, and supporting the less privileged. Her efforts in promoting equality, anti-discrimination, and social justice have made her a respected figure in her community. As an author, Mrs. Zainab Melvina Omoyinmi brings a wealth of knowledge and experience to the discussion on good governance and democratic principles. Her contributions to this book are grounded in her extensive background in social work and community development, offering valuable insights into the importance of safeguarding human rights and promoting social equity.

A devoted mother of four, Mrs. Omoyinmi balances her professional and advocacy roles with her responsibilities to her family, embodying the principles of compassion and service that she champions in her work.

Author's Bio: Hassan Augustine Turay (HAT) is a seasoned professional with over two decades of experience in Human Resource Management and Development, complemented by a strong background in general administration. His career has been marked by a commitment to fostering effective team dynamics, leading large organizations, and driving professional development initiatives. Known for his excellent communication, interpersonal, and leadership skills, Hassan has consistently demonstrated his ability to manage complex tasks, prioritize workloads, and thrive in high-pressure environments.

Hassan holds a professional qualification in Human Resource Management & Development, which has equipped him with the knowledge and skills necessary to excel in various roles within the field.

His expertise extends to learning and development, team leadership, and staff supervision, areas where he has made significant contributions in both corporate and community settings.

With a passion for problem-solving and a proactive approach to challenges, Hassan is recognized for his ability to build rapport within teams and adapt to changing environments. His enthusiasm, coupled with a can-do attitude and a good sense of humor, makes him a respected and effective leader.

As an author, Hassan Augustine Turay brings a wealth of practical experience to the discourse on good governance and democratic principles. His contributions to this book are informed by his extensive background in human resource management and his commitment to fostering organizational integrity and ethical leadership. Hassan's insights offer valuable perspectives on the role of effective administration and leadership in promoting democratic values and ensuring the success of governance structures.

Through his writing, Hassan aims to share his knowledge and experiences to contribute to the development of a more just and equitable society in Sierra Leone, where good governance and democratic principles are upheld and respected.

Author's Bio: Prof. M.Y. Bangura, widely known by his pen name, Mye Bangura, brings over 30 years of experience in education, consultancy, and development. As the Project Coordinator for the Sierra Leone Advocacy Movement (SLAM-Global), he plays a key role in promoting democratic governance and social development in Sierra Leone. Born to parents from the Port Loko and Tonkolili districts in northern Sierra Leone, Prof. Bangura's life and career reflect a deep commitment to education and public service.

Prof. Bangura's professional journey began in the education sector, where he taught in primary and secondary schools in Sierra Leone. His passion for teaching and leadership saw him rise to significant roles, including Director of Research and Development at Every Nations Polytechnic College (affiliated with EBK University) and the first Principal of the Civil Service Training College (CSTC) of Sierra Leone, a position he held for a decade (2012-2022). During his tenure, he revived and restructured the CSTC, using his project management expertise to develop new initiatives and curricula that transformed the institution into a vital training ground for public servants.

A highly respected academic, Prof. Bangura has been instrumental in designing numerous educational programs that have shaped the landscape of higher education in Sierra Leone. His contributions include establishing diploma and certificate courses at IAMTECH, EBK University, and Fatima Institute (now University of Makeni), as well as pioneering the first postgraduate program (MA in Sustainable Development) at the University of Makeni. His efforts have been pivotal in the transformation of these institutions into recognized universities.

Internationally, Prof. Bangura has served in various academic and leadership roles in the UK, including Vice Principal of A-Mark International College in London and Head of Business Communications & Development at St. Martins Business School. He has also worked as an international consultant, contributing to projects in Europe, America, and Africa. His expertise in project development, research, and education has made him a sought-after consultant in both the public and private sectors.

In addition to his educational work, Prof. Bangura has made significant contributions to agriculture and food security through the Kampala Agricultural Farm Project, which he established in 2018. His work in this field focuses on sustainable farming practices and improving food production in Sierra Leone.

As an author of this book, Prof. Bangura draws on his extensive experience in education, governance, and development to offer insights into how democratic principles can be strengthened in Sierra Leone. His commitment to improving standards of living and promoting good governance makes him a key voice in the ongoing development of Sierra Leone's future.

Author's Bio: Foday John Turay is a dedicated advocate for the intersection of cultures and ideas, with a rich background that spans both the Global South and North. Born and educated in Sierra Leone, Foday's early years were marked by extensive travel throughout the country, experiences that instilled in him a deep understanding of its diverse cultures and geography. His passion for education and social sciences led him to pursue further studies in the United States, where he gained graduate and postgraduate qualifications, broadening his worldview and reinforcing his commitment to bridging the gap between Africa and the Global North.

Throughout his varied career, Foday has worn many hats. He is a veteran of the Sierra Leone Army, an educator who has taught at both the high school and teacher-training college levels, a psychiatric counselor, and a travel consultant. His diverse professional experiences reflect his wide-ranging interests in news, current affairs, geography, and the social sciences. A self-described "newsaholic" with an eclectic appetite for knowledge, Foday's work is driven by a desire to apply his insights to real-world challenges, particularly in the context of Sierra Leone's development.

As an author, Foday brings a unique perspective shaped by his experiences in both Sierra Leone and the United States. His deep understanding of these distinct cultures allows him to serve as an intermediary, fostering dialogue and cooperation between Africa and the Global North. His contributions to this book are informed by his extensive travels and his comparative view of life in both regions, offering readers valuable insights into the benefits and challenges presented by each.

Through his work with the Sierra Leone Advocacy Movement (SLAM) and its publications, Foday is committed to improving the standard of living and promoting democracy in Sierra Leone and across Africa. He sees his role not only as an educator and advocate but also as an ambassador of goodwill, working to build stronger connections between Sierra Leone and the United States.

Foday John Turay's interests include news and current affairs, African literature and history, traveling, and sports. These passions continue to inform his work as he strives to make a meaningful impact on the lives of others through his writing and advocacy.

Author's Bio: Born in Mabora, Sierra Leone, Mrs. Fatima Sesay is a dedicated community leader and entrepreneur with decades of experience in service and leadership. From an early age, Fatima demonstrated her natural leadership abilities by taking on significant responsibilities, including caring for her younger siblings long before reaching adulthood. These early experiences shaped her commitment to supporting and uplifting those around her, a theme that has carried through to her various ventures.

Mrs. Sesay has successfully managed several initiatives that bring lasting benefits to her community. She has been instrumental in establishing systems to care for orphans, resolving family disputes, and creating job opportunities through business ventures, such as her latest project—a farm back in Sierra Leone that employs many local workers. Her efforts reflect her deep commitment to improving the lives of others, particularly those in her native community.

In addition to her entrepreneurial work, Mrs. Sesay is a qualified nurse, a profession that further highlights her nurturing and service-oriented nature.

As a successful mother, she is also known for her steadfast support of her family and her ability to offer solutions in the face of challenges. Her resilience and ability to navigate complex situations make her a respected figure in both her personal and professional circles.

As an author of this book, Mrs. Fatima Sesay brings a wealth of practical experience in leadership, community development, and healthcare. Her story is one of resilience and dedication, offering readers valuable insights into how grassroots efforts can lead to meaningful change. Through her writing, she aims to inspire others to act in their communities and contribute to the ongoing development of Sierra Leone.

Author's Bio: Abdul Fonti was born in Foredugu, a vibrant community within the Romende Chiefdom of the Karene District in the Northern Province of Sierra Leone. This region, known for its rich cultural heritage and close-knit communities, provided the foundation for Abdul's lifelong commitment to social justice and youth empowerment.

Hailing from a family deeply rooted in the principles of fairness, equity, and service to others, Abdul's early life was shaped by the values of his ancestors, who were known for their leadership in advocating for the welfare of their community. Inspired by these values, Abdul has spent decades working tirelessly to empower the youth of Sierra Leone, believing that the nation's future rests in their hands. His youth empowerment efforts have provided opportunities for education and personal development and have inspired a generation of young leaders who are now actively contributing to their communities.

Abdul's advocacy extends beyond youth empowerment; he is passionate about democratic principles, particularly in transparency and equitable wealth distribution. He firmly believes that Sierra Leone's wealth should be shared in a manner that benefits all its citizens, not just a privileged few. This belief has driven his involvement in various initiatives aimed at promoting good governance and ensuring that Sierra Leone's resources are managed in a way that fosters national development and social justice.

In pursuit of his advocacy goals, Abdul Fonti has also been an active voice in political debates within Sierra Leone and the diaspora. His move to Houston, Texas, USA, has not diminished his commitment to his homeland. Instead, it has provided him with a broader platform to engage with the Sierra Leonean diaspora and international communities, advocating for democratic reforms and the equitable distribution of resources.

Despite his deep involvement in social and political issues, Abdul remains a humble individual who finds joy in life's simple pleasures.

He is an avid soccer enthusiast, a sport that he believes teaches valuable lessons in teamwork, strategy, and perseverance—equally essential qualities in leadership and governance.

Abdul Fonti's life is a testament to the power of dedicated advocacy and the impact one individual can have on the lives of many. His work continues to inspire those around him, as he remains committed to the ideals of transparency, justice, and the empowerment of Sierra Leone's youth. His journey from Foredugu to Houston is not just a personal narrative but a beacon of hope and a call to action for all Sierra Leoneans, both at home and abroad, to engage in the collective effort of nation-building.

Author's Bio: Mariatu Kamara (née Bangura) was born and raised in Freetown, Sierra Leone, a city rich in history and culture. Growing up in this vibrant environment, she developed a deep love for education and community service from an early age. Her academic journey began at St. Joseph's Convent Brookfields, one of Freetown's prestigious schools, where she laid a strong academic foundation and cultivated a passion for discipline and perseverance.

Mariatu continued her education at the Collegiate Secondary School, where she nurtured her interest in business and finance. This passion led her to pursue higher education at the Institute of Public Administration and Management (IPAM), where she earned a Bachelor of Science degree in Business Administration and Finance. Her academic achievements set the stage for a distinguished career in finance.

Starting her professional journey as a Finance Assistant at Fourah Bay College, University of Sierra Leone, Mariatu served diligently for 11 years. Her expertise in finance and administration significantly contributed to the institution's financial operations, reinforcing her reputation as a capable and dedicated finance professional. Her work at Fourah Bay College deepened her commitment to education and its critical role in national development.

Seeking new challenges, Mariatu relocated to the United States, where she furthered her education by obtaining a master's degree from Abilene Christian University in Texas. This accomplishment enriched her knowledge and opened new professional opportunities in the U.S., beginning with a position at the City of Dallas. Here, she gained valuable insights into public administration and financial management in a large metropolitan setting.

Currently, Mariatu plays a pivotal role with the Department of Family and Protective Services in Dallas, Texas, where she contributes to safeguarding the well-being of children and families. Her work ensures that resources are allocated efficiently to protect society's most vulnerable members, reflecting her dedication to public service and community welfare.

Throughout her journey, Mariatu has remained deeply connected to her Sierra Leonean roots, drawing inspiration from her upbringing in Freetown. Her life's work, both in Sierra Leone and the United States, reflects a commitment to using her skills to serve others, particularly in education, finance, and social welfare. Mariatu's story is a testament to resilience, dedication, and the transformative power of education, inspiring young women in Sierra Leone and beyond to pursue excellence and lifelong learning.

Author's Bio: Marie H. Kamara is a dedicated community leader, healthcare professional, and advocate for social development. Born and raised in Freetown, Sierra Leone, Marie's roots trace back to the Northern Province, specifically Kambia District, where her family hails from. With a passion for helping others and giving back to her community, she has made significant contributions both in Sierra Leone and in the diaspora.

Currently residing in the Netherlands, Marie works as a registered nurse, a role that reflects her commitment to service and care for others.

Her professional background in healthcare has equipped her with the skills and compassion necessary to advocate for those in need. In addition to her nursing career, Marie is also a respected leader within the Sierra Leonean community abroad. She serves as the CEO of Global Friends of APC, a grassroots organization that mobilizes support for the All People's Congress (APC) party, and she holds the position of Deputy Chairlady of the Holland District.

Marie is deeply committed to supporting local communities in Sierra Leone, particularly in her native Kambia District. She is the founder and CEO of the Africa Aid Initiative, a nonprofit organization that focuses on empowering local communities in the Masogbala Chiefdom of Kambia. Through this initiative, she has worked tirelessly to provide much-needed resources and support in areas such as healthcare, education, and social welfare, helping to improve the lives of many in the region.

As an author of this book, Marie brings a wealth of experience from her work in both healthcare and community development. Her dedication to fostering connections between the Sierra Leonean diaspora and communities back home is a testament to her belief in the power of collective action for social change. Through her writing, Marie aims to inspire others to contribute to the development of Sierra Leone by promoting good governance, democratic principles, and community empowerment.

Printed in Great Britain
by Amazon